Engineering and Society: Working Towards Social Justice Part II: Engineering: Decisions in the 21st Century

Engineering and Society: Working Towards Social Justice
Part II: Engineering: Decisions in the 21st Century
George Catalano

ISBN: 978-3-031-79951-8   paperback
ISBN: 978-3-031-79952-5   ebook

DOI 10.1007/978-3-031-79952-5

A Publication in the Springer series
SYNTHESIS LECTURES ON ENGINEERS, TECHNOLOGY AND SOCIETY

Lecture #9
Series Editor: Caroline Baillie, *University of Western Australia*

Series ISSN
Synthesis Lectures on Engineers, Technology and Society
Print 1933-3633   Electronic 1933-3461

Drawings © 2009 by Z*qhygeom

Synthesis Lectures on Engineers, Technology and Society

Editor
Caroline Baillie, *University of Western Australia*

Engineering Ethics: Peace, Justice, and the Earth
George D. Catalano
2006

Engineering and Society: Working Towards Social Justice Part II: Engineering: Decisions in the 21st Century

George Catalano
State University of New York at Binghamton

SYNTHESIS LECTURES ON ENGINEERS, TECHNOLOGY AND SOCIETY #9

ABSTRACT

Part II: Engineering: Decisions in the 21st Century Engineers work in an increasingly complex entanglement of ideas, people, cultures, technology, systems and environments. Today, decisions made by engineers often have serious implications for not only their clients but for society as a whole and the natural world. Such decisions may potentially influence cultures, ways of living, as well as alter ecosystems which are in delicate balance. In order to make appropriate decisions and to co-create ideas and innovations within and among the complex networks of communities which currently exist and are be shaped by our decisions, we need to regain our place as professionals, to realise the significance of our work and to take responsibility in a much deeper sense. Engineers must develop the 'ability to respond' to emerging needs of all people, across all cultures. To do this requires insights and knowledge which are at present largely within the domain of the social and political sciences but which needs to be shared with our students in ways which are meaningful and relevant to engineering. This book attempts to do just that. In Part 1 Baillie introduces ideas associated with the ways in which engineers relate to the communities in which they work. Drawing on scholarship from science and technology studies, globalisation and development studies, as well as work in science communication and dialogue, this introductory text sets the scene for an engineering community which engages with the public. In Part 2 Catalano frames the thinking processes necessary to create ethical and just decisions in engineering, to understand the implications of our current decision making processes and think about ways in which we might adapt these to become more socially just in the future. In Part 3 Baillie and Catalano have provided case studies of everyday issues such as water, garbage and alarm clocks, to help us consider how we might see through the lenses of our new knowledge from Parts 1 and 2 and apply this to our every day existence as engineers.

KEYWORDS

engineering and society, social justice, ethics, engineering education, globalisation, public dialogue with engineering, engineering development, engineering studies

In memory of beloved Francesca Clarisa
who brought unconditional love to my life
and to all those knew her.
I miss her presence and wise counsel
today as I do all days.

Contents

CHAPTER 1

Making Decisions in the 21st Century

1.1 INTRODUCTION

We all strive for safety, prosperity, comfort, long-life, and dull-ness. The deer strives with his supple legs, the cowman with trap and poison, the statesmen with his pen, the most of us with machines, votes and dollars, but it all comes to the same thing: peace in our time.

Aldo Leopold, A Sand County Almanac[a].

[a]Aldo Leopold (1887-1948) is considered the father of wildlife ecology. He was a renowned scientist and scholar, exceptional teacher, philosopher, and gifted writer. It is for his book, A Sand County Almanac, that Leopold is best known by millions of people around the globe. Published in 1949, shortly after Leopold's death, A Sand County Almanac is a combination of natural history, scene painting with words, and philosophy. It is perhaps best known for the following quote, which defines his land ethic: "A thing is right when it tends to preserve the integrity, stability, and beauty of the biotic community. It is wrong when it tends otherwise."

Figure 1.1: Leopold.

Engineers make choices in nearly all aspects of their work. This text aims to provide you with the ability to make choices in an informed, reflective way. Here we are not concerned with choices for lunch menus or for after dinner engagements but rather making decisions concerning some of the most challenging dilemmas you will likely encounter throughout your careers. As we move farther into the 21st century, engineers will become more directly involved in issues of conflict, development and environmental sustainability. We shall confront those issues head on and offer a variety of frames of reference whereby you as an engineer can make choices whether the issue deals with some small aspect of a design or to work on a project at all. The frames of reference for decision-making include traditional approaches used in engineering throughout the modern era as well as new and exciting ideas which have just recently been applied to the professions.

Engineering is a profession with an important ethical dimension (Davis, 1998). First, we shall explore the significance of ethics for our profession, paying careful attention to the importance of codes of ethics adopted by various engineering societies. Next, as a mechanism for introduction of the complexity of issues that you may confront as engineers in the 21st century, we shall introduce

> *Can we talk of integration until there is integration of hearts and minds? Unless you have this, you only have a physical presence, and the walls between us are as high as the mountain range.*
>
> *Chief Dan George[a].*

[a]Chief Dan George (1899-1981) is best known for his writing and acting careers. His most famous roles were as the aging Chief, opposite Dustin Hoffman, in "Little Big Man," and the aging Chief opposite Clint Eastwood in "Josey Wales." He used his writing and media roles to try to give a more accurate depiction of American Indian beliefs and values.

Figure 1.2: Chief Dan George.

and discuss three cases: conflict and peace, poverty and development and environmental degradation and sustainability. These cases set the stage for our in-depth look at engineering ethics as it has been applied in the past and presently and offer you several new ideas for handling even more complex situations in the future. The new approaches include an ethic based on freedom, one on chaos, one on a morally deep world view, one consistent with a global ethic and lastly, an engineering ethic based on love.

Our goal is to enable you to consider the various issues from a wide-range of perspectives and thus ultimately be able to make your choice consistent with your deepest held values and convictions. In the spirit of the Diggers[1] from the 1960's, we are offering new 'frames of reference' from which you can consider your decisions. The Diggers focused on promoting a new vision of society free from many of the trappings of private property, materialism and consumerism. We hope to offer you a new vision of engineering which takes into account many of the elements of our society and our planet which have been historically ignored. Let us get started!

1.2   INTRODUCTORY NOTE

As you progress through the text, you will be routinely confronted with the following two items: first, we hope you come to recognize and identify with this engineering student as she progresses through the course material; and secondly, accompanying this sketch will be a 'Challenge Box' which

---

[1] The Diggers were one of the legendary groups in San Francisco's Haight-Ashbury, one of the world-wide epicenters of the Sixties Counterculture which fundamentally changed American and world culture. The Diggers took their name from the original English Diggers (1649-50) who had promulgated a vision of society free from private property, and all forms of buying and selling. The San Francisco Diggers evolved out of two Radical traditions that thrived in the SF Bay Area in the mid-1960s: the bohemian/underground art/theater scene, and the New Left/civil rights/peace movement.

will provide you with questions that will challenge your understanding of the material presented and also call on you to make connections to your own life and experiences in the world.

> **Challenge Box:** Describe the universe. Give two examples!

1.3   USING ETHICS CASES

The use of case studies has a distinguished history in law and business schools, and it has been very successful in the more recent emergence of medical ethics as an area of study. One of the major reasons the use of case studies is so successful is that ethical inquiry begins with problems that professionals can expect to have to face. This is in contrast to beginning at a highly theoretical level and only later considering how rather general principles and rules might apply to actual situations. By using realistic cases, students can immediately appreciate the relevance and importance of giving serious thought to ethics. Careful reflection on the cases will itself suggest the need for moving to a more theoretical level.

Purposefully and deliberately, there is a significant difference in the types of cases that we have chosen for inclusion in the present text. Many outstanding texts and other resources in engineering ethics already exist[2] [3]. Most tend to focus on either micro- and macro-ethics or a second approach described as the "tragic case" methodology (Liaschenko et al., 2002). Micro-ethics focuses upon issues related to professionalism such as integrity, honesty and reliability, risk and safety and responsibilities as an employee. Macro-ethics broadens the coverage to include issues related to the impact of engineering on the environment or the societal context of engineering. An example of a typical case in micro-ethics would be one provided by Harris et al. (2008) in which a manufacturing

---

[2] Online Ethics Center at National Academy of Engineering: As part of the new Center for Engineering, Ethics, and Society at the National Academy of Engineering, the mission of the Online Ethics Center at the National Academy of Engineering is to provide engineers and engineering students with resources for understanding and addressing ethically significant problems that arise in their work, and to serve those who are promoting learning and advancing the understanding of responsible research and practice in engineering. (www.onlineethics.org)

[3] Science and Engineering Ethics is a multi-disciplinary journal that explores ethical issues of direct concern to scientists and engineers. Coverage encompasses professional education, standards and ethics in research and practice, extending to the effects of innovation on society at large. (www.springer.com/philosophy/ethics/journal/11948)

engineer is invited by a vendor to play golf at an exclusive, private club. An example of a macro-ethics case has been offered by Herkert (2008) in his critique of the ethical responsibilities of engineering with respect to global climate change.

According to Allenby (2005), these two major approaches suggests a third. First, Allenby makes a differentiation between "micro-ethics," or ethics at the individual level, and "macro-ethics," or ethics at a group or professional level. Second, he asserts "a principal result of the Industrial Revolution and concomitant demographic, economic, technological and social changes is a planet where the dynamics of most major natural systems are increasingly shaped by human activity." The journal *Nature* suggests that we have now entered the *Anthropogenic Age* or the *Age of Man* (2003). In conclusion, Allenby argues that there is a gap between our current ethical systems and the world we have created, that it is a major gap, and that addressing it requires the serious and integrated efforts of the technical, scientific, and philosophic communities. It is that very gap which has served for the catalyst for the approach we have taken in this text.

The Worldwatch Institute has identified the three most important issues for the 21st century, which are the issue of conflict vs. peace and security, the issue of poverty vs. development and the issue of environmental deterioration vs. sustainability (Worldwatch, 2007). Accordingly to begin our discussion of making engineering decisions we introduce a set of cases which link to the Worldwatch Institutes identified issues. The case involving conflict vs. peace explores the ethical issues related to landmines. Considering poverty vs. development, we shall highlight the aftermath of Hurricane Katrina on the residents of New Orleans while our example for reflecting upon the ethical responsibilities associated with the environment centers on the rapidly disappearing bumblebee colonies. These three cases then set the stage for what is to follow.

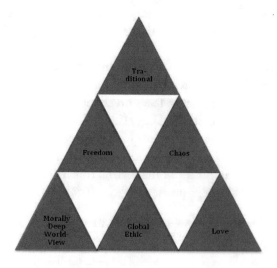

Figure 1.3: Approaches to solving ethical problems.

Figure 1.4: Specific ethical cases to be explored using various applied ethical approaches.

1.4   ENGINEERING PROFESSION

According to the Accreditation Board for Engineering and Technology[4], engineering is the profession in which a knowledge of the mathematical or physical sciences gained by study, experience and practice is applied with judgment to develop ways to utilize, economically, the materials and forces of nature for the benefit of mankind. Like the fields of medicine and law, engineering is considered a profession. But what is a profession, and how is it characterized? In a very general sense, a person's occupation is often referred to as his or her profession. In a stricter sense, however, a profession is characterized by special qualifying characteristics. According to Webster's Unabridged Dictionary (2005), a profession requires specialized knowledge and often requires considerable preparation in skills, methods, and principles[5]. Professions also maintain high standards of achievement and conduct, encourage life long learning, and engage in public service.

Engineers and scientists take many mathematics, chemistry, and physical courses in common in their undergraduate programs. However, their paths diverge as they begin to take courses in their majors, with engineering majors taking courses that are more applied and science majors taking

---

[4] ABET, Inc., is the recognized U.S. accrediting agency of college and university programs in applied science, computing, engineering, and technology. ABET was established in 1932 and now consists of 28 different professional and technical societies representing the fields of applied science, computing, engineering, and technology.

[5] A profession is an occupation, vocation or career where specialized knowledge of a subject, field, or science is applied. It is usually applied to occupations that involve prolonged academic training and a formal qualification. Professional activity involves systematic knowledge and proficiency. Professions are usually regulated by professional bodies that often develop and offer examinations to assess competence, act as a licensing authority for practitioners, and enforce adherence to an ethical code of practice. (http://en.wikipedia.org/wiki/Profession)

more courses in the basic sciences, being more interested in discovering knowledge with less regard to its immediate application.

On the job, engineers create tangible devices or systems such as computers, engines, and other products that perform a function. Engineers may also be involved with the process of designing systems such as satellite communications systems and robotics control systems, or with the development of a practical solution to a meet human need. Scientists also solve problems and may develop new products, but the emphasis is on the science rather than the product.

Challenge Box: Think of five professions in addition to engineering. Compare and contrast the professions in terms of their actual and potential impacts both positive and negative on society and on the environment.

CHAPTER 2

Ethics

Figure 2.1: Ethical dilemmas.

If it is not right do not do it; if it is not true do not say it.

Meditate often upon the bond of all the Universe and their mutual relationship. For all things are in a way woven together and all are because of this dear to one another; for these follow in order one upon another because of the stress movement and common spirit and unification of the matter.

Marcus Aurelius, **Meditations** (Aurelius, 2002).

Figure 2.2: Aurelius.

2.1   INTRODUCTION

Mrs. Applebee pulled down the shade
And retired to her cozy bed
Officer Warner, round the corner
Sippin' coffee and a makin' time
He never heard the sounds of violence
Emanating from the crime
April Stark was in the park
Conducting business from her usual spot
She was witness to what happened
But conveniently forgot
Funky Wheeler, he's no squealer
He flipped his quarter high into the air
He saw their faces, but told detectives,
"I don't know nothing Jack nor do I care."
Mary's mother asked the question
"Didn't anybody hear?
When my daughter cried out "Help me!"
Didn't anybody care!?"
All's quiet on West 23rd
Nobody saw, nobody heard
All's quiet on West 23rd
Nobody saw, nobody heard
Lyrics by Julie Budd

[Reprinted with permission from the March 27, 1964 *New York Times*. Copyright© 1964 by the New York Times Co.]
37 Who Saw Murder Didn't Call the Police Apathy at Stabbing of Queens Woman Shocks Inspector
By *Martin Gansberg*
For more than half an hour thirty- eight respectable, law-abiding citizens in Queens watched a killer stalk and stab a woman in three separate attacks in Kew Gardens. Twice the sound of their voices and the sudden glow of their bedroom lights interrupted him and frightened him off. Each time he returned, sought her out and stabbed her again. Not one person telephoned the police during the assault; one witness called after the woman was dead.

In March, 1964, a New York City woman named Catherine "Kitty" Genovese was raped and stabbed to death as she returned home from work late at night. According to a newspaper report published shortly thereafter, her neighbors looked on from their bedroom windows for over 30 minutes as the assailant beat her, stabbed her, left her and returned to repeat the process two more times until she died. No one lifted a phone to call the police; no one shouted at the criminal; no one went to the aid of the young woman. Finally, a 70 year old woman called the police. Police officers arrived two minutes later but by that time the young woman was dead. For the next hour while authorities awaited the arrival of a ambulance, only one other woman came out to testify. At that time, many residents came out of their homes. The New York Times reporter at the scene of the crime reported that 38 people had witnessed some or all of the attack, which took place in two or three distinct episodes over a period of about a half hour. Notwithstanding the large number of witnesses, no one did anything to stop the attack; no one even reported it to the police until the woman was already dead. When asked why no one did anything, the responses ranged from "I don't know" to "I was tired" to "I was scared." Although the murder itself was tragic, the nation was even more outraged that so many people who could have helped seemingly displayed callous indifference.

The Genovese incident is one of the most highly discussed examples of onlookers remaining indifferent to the plight of a victim of a violent crime. Recently, another incident in Cleveland, Ohio, is a reminder that such occurrences are still commonplace. On June 28, 2008, a group of teenagers beat a homeless man to death as passers-by slowed to watch the attack, some of which was caught on video tape. The homeless man, Anthony Waters, 42, suffered a lacerated spleen and broken ribs and died at a nearby Cleveland hospital later that evening. The tape showed multiple cars slowing to watch the teens attack Waters. No one stopped or offered any assistance or called the police. The attackers appeared to be between the ages of 14 and 17, robbed Mr. Waters of a music player and headphones.

Such incidents are not confined to the U.S.; Italian newspapers, an archbishop and civil liberties campaigners expressed shock and revulsion after photographs were published of sunbathers apparently enjoying a day at the beach just meters from where the bodies of two drowned Roma[1] girls were laid out on the sand. Italian news agency ANSA reported that the incident had occurred on Saturday at the beach of Torregaveta, west of Naples, southern Italy, where the two girls had earlier been swimming in the sea with two other Roma girls. Reports said they had gone to the beach to beg and sell trinkets.

Local news reports said the four girls found themselves in trouble amid fierce waves and strong currents. Emergency services responded 10 minutes after a distress call was made from the beach and two lifeguards attended the girls upon hearing their screams. Two of them were pulled to safety but rescuers failed to reach the other two in time to save them. The website of the Archbishop of

---

[1] There are more than twelve million Roma located in many countries around the world. There is no way to obtain an exact number since they are not recorded on most official census counts. Many Roma themselves do not admit to their true ethnic origins for economic and social reasons. The Roma are a distinct ethnic minority, distinguished at least by Rom blood and the Romani, or Romanes, language, whose origins began on the Indian subcontinent over one thousand years ago. No one knows for certain why the original Roma began their great wandering from India to Europe and beyond, but they have dispersed worldwide, despite persecution and oppression through the centuries.

Naples said the girls were cousins named Violetta and Cristina, aged 12 and 13. Their bodies were eventually laid out on the sand under beach towels to await collection by police. Photographs show sunbathers in bikinis and swimming trunks sitting close to where the girls' feet can be seen poking out from under the towels concealing their bodies. A photographer who took photos at the scene told CNN the mood among sunbathers had been one of indifference.

Challenge Box: Suppose you were witness to a crime as horrific as the Genovese or the Waters case. What would you have done if: (1) the victim was a complete stranger; (2) the victim was a neighbor you recognized; (3) a friend; (4) a member of your immediate family; or (5) a neighborhood dog or cat? Are your answers different? Why or why not? How did you decide what you would do? Did you use any principles? Suppose there were two assailants and each was heavily armed? Would that change your responses? Why or why not?

Or suppose you were sun bathing on an exclusive beach in the French Riviera or some other exotic place. That very day you were accosted by a roving band of gypsies and they managed to have stolen your wallet/purse with all your traveler's checks. You were able to quickly get a refund and continue on with your vacation. While visiting the nearby beach, sipping a delicious alcoholic concoction and working on your tan, you notice that the several members of the same band of gypsies had tried to swim and due to strong currents had drown and their bodies were laid out on the beach close to you, covered by a blanket. How would you feel? What would you do if anything?

The Genovese, Waters and Roma children tragedies suggest a range of questions for each of us. Who should I care about? Who is my neighbor? What should the neighbors or passers-by have done in these cases? What would each of us do in a similar situation? Do I have a moral obligation to try to help unfortunate victims such as Ms. Genovese or Mr. Waters or gypsies? What does it mean to be a moral person in the world? Why should I be a moral person? What is morality anyway? Why do we even need morality? Is it in my interest to be a moral person? What is the basis of morality? What does it mean to be moral when it involves personal sacrifice? Such questions serve as an entry into an aspect of philosophy which deals with the question how should we live. That aspect, moral philosophy, is also referred to as ethics.

2.2   ENGINEERING AND CODES OF ETHICS

Engineering is the discipline of acquiring and applying scientific and technical knowledge to the design, analysis, and/or construction of works for practical purposes. ABET defines engineering as: "The creative application of scientific principles to design or develop structures, machines, apparatus, or manufacturing processes, or works utilizing them singly or in combination; or to construct or operate the same with full cognizance of their design; or to forecast their behavior under specific operating conditions; all as respects an intended function, economics of operation and safety to life and property" (Holmes, 2002). The broad discipline of engineering encompasses a range of specialized sub-disciplines that focus on the issues associated with developing a specific kind of product, or using a specific type of technology.

Engineering, similarly to law and medicine, is a profession, which means that it is an occupation, vocation or career where specialized knowledge of a subject, field, or science is applied (Wikipedia, 2008). It is usually applied to occupations that involve prolonged academic training and a formal qualification. It is axiomatic that "professional activity involves systematic knowledge and proficiency" (Kashner, 2005). Professional bodies that may set examinations of competence, act as a licensing authority for practitioners, and enforce adherence to an ethical code of practice usually regulate professions.

Engineering ethics is the field of applied ethics, which examines and sets standards for engineers' obligations to the public, their clients, employers and the profession (Petroski, 1985). Engineering does not have a single uniform system, or standard, of ethical conduct across the entire profession. Ethical approaches vary somewhat by discipline and jurisdiction, but are most influenced by whether the engineers are independently providing professional services to clients, or the public if employed in government service; or if they are employees of an enterprise creating products for sale.

Engineers decide to become members of the profession of their own free will; thus, one key aspect of the codes of ethics or conduct as described throughout the various sub-disciplines of engineering is that they have been set forth and accepted voluntarily as explained by Davis (1999). A second point offered by Davis is that ethical codes by their very nature are moral. One cannot speak of immoral ethical codes or principles; there is no such phenomenon as a Nazi code of ethics.

Engineering ethics is an example of the category in ethics referred as applied ethics. Applied ethics, which typically deals with controversial moral problems, is one of the key divisions within the study of ethics, which also includes descriptive morality (i.e., the actual beliefs, customs, principles and practices of people and cultures) and moral philosophy or ethical theory (i.e., the systematic effort to understand moral concepts and justify moral principles). A legitimate question one might ask is why we as engineers should be concerned with moral philosophy? Is not it enough to simply know and embrace our codes of ethics as engineers? Perhaps, but we would argue that such may not be the case at all. As argued by Pojman and Fieser (2009), "Theory without application is sterile and useless, but action without a theoretical perspective is blind." In today's ever shrinking planet, the need for an awareness of, an appreciation for and, ideally, an understanding of the diverse beliefs, customs, and principles of the various cultures and societies throughout the world appears to be more important today than ever before. As the inexorable forces of globalism bring us even closer, that need is going to become even more important in the future.

2.3   ETHICS AND THE ETHICAL SEQUENCE

Fundamentally, an ethic from a philosophical perspective is a differentiation of social from anti-social behavior. From the perspective of a systems engineering, a different kind of perspective for example, an ethic can be seen as a limitation of freedom of action in support of the advancement or health of the system as a whole. Contrast these two views with that of an ecologist who may imagine an ethic as limitation on freedom of action in the struggle for existence. Whatever the discipline, an ethic has at its origin the tendency for interdependent individuals or groups to evolve modes of co-operation. A sociologist may refer to the holistic structure as a community; an ecologist may refer to it as an eco-system, and a political scientist may refer to it as a government. And what would be the equivalent for an engineer? Who and what matters for the engineering professional? The answers to such questions are the focus of this present text.

Rather than describing an ethic whether pure or applied, perhaps, it is more appropriate to consider an ethical sequence. Such a sequence would move from the simplest of ethics, which deal with the relationships among individuals to the relations between the individuals and the various communities of which that individual is a part to the interdependency of humankind to the environment. The *Ten Commandments* or *Mosaic Decalogue* (Catholic Encyclopedia, 2008) is an example of an ethic, which focused in part on relationships among individuals[2]. Gandhi sought to transform such restrictions from the negative to a positive construction. In contrast to the *Mosaic Decalogue* of negatives, Gandhi (Singh, 2004) provided seven root causes of unfairness and injustice, all consisting of volitional human activities in the absence of socially redeeming moral content[3].

The integration of an individual into a greater community or society is the subject of the ethic of reciprocity or *The Golden Rule* (Teaching Values, 2008), a fundamental moral principle that simply means, "treat others as you would like to be treated." *The Golden Rule* may be the most basic of all ethics. The rule can be found in various forms in a wide array of the earth's wisdom traditions. According to the Greeks, "What you wish your neighbors to be to you, such be also to them" (Graves,

---

[2] The Ten Commandments according to many scholars serve a dual purpose; they form a covenant between God and his people, and serve as a moral law, or code, by which his people are to live by. To achieve the full observance of the Ten Commandments, an elaborate system, known as Mosaic Law, was put into place. This Mosaic Law involved a vast legal system of Israel, civil, criminal, judicial, and ecclesiastical framed after the Decalogue. The Mosaic Law was to be a temporary experiment, while the Decalogue was to be permanent.

[3] Gandhi's Seven Root Causes include: wealth without work; pleasure without conscience; knowledge without character; commerce without morality; science without humanity; worship without sacrifice; and politics without principles.

[4] Mahatma Gandhi (1869—1948) lead India to independence using non-violent means. Indians refer to him as Mahatma Gandhi which means great soul. Martin Luther King later used these same techniques to fight for black civil rights in the USA.

Figure 2.3: Mahatma Gandhi.[4].

Figure 2.4: The 14th Dalai Lama.

1876). In the modern world, the Dalai Lama has stated: "If you want others to be happy, practice compassion. If you want to be happy, practice compassion"[5].

The notion of a sequence moving towards greater complexity parallels the evolutionary process that characterizes the unfolding of the universe from single particle to immense gas clouds to the formation of stars and planets and ultimately life.

It is not an oversimplification to state that all ethics, at least those developed so far, have as their foundation the proposition that the individual is a member of a community of interdependent parts. The identification of the various members of the community and their rejection or inclusion into the dialog may yield very different answers when confronted with an ethical dilemma whether it is in engineering, some other discipline or life itself.

Challenge Box: Were you surprised that all of the major world wisdom traditions include some form of the Golden Rule? Explain. What are the implications for you?
Do you believe our sense of ethical responsibility is evolving? Will it reach a steady state? What would such a steady state look like? Or will it continue to grow inexorably? What are the implications for you as you begin your career?

[5] His Holiness the XIVth Dalai Lama, Tenzin Gyatso, is the spiritual and temporal leader of the Tibetan people. The Dalai Lamas are the manifestations of the Bodhisattva of Compassion, who chose to reincarnate to serve the people. Dalai Lama means Ocean of Wisdom.

CHAPTER 3

Landmines and the War in Iraq

It is no longer a choice, my friends, between violence and nonviolence. It is either nonviolence or nonexistence. And the alternative to disarmament, the alternative to greater suspension of nuclear tests, the alternative to strengthening the United Nations and thereby disarming the whole world, may well be a civilization plunged into the abyss of annihilation, and our earthly habitat would be transformed into an inferno that even the mind of Dante could not imagine.

Martin Luther King, Jr. (King et al., 1990).

Figure 3.1: Martin Luther King.

I'm fed up to the ears with old men dreaming up wars for young man to die in.

George McGovern.

Figure 3.2: George McGovern.

A recent report documents the effects of the 2003 war in Iraq and the ensuing period after the fall of the Baathist regime on health, the health system, and the health system reconstruction. The World Health Organization (WHO) defines health as follows: "a state of complete physical, mental and social well-being and not merely the absence of disease or infirmity." The WHO argues that the impact of war and violence, in addition to the deaths and injuries due to weaponry, also includes the often greater longer-term suffering linked with damage to the essential infrastructure, a poorly functioning health system and the failure of relief and reconstruction efforts (Panch, 2004).

It may be useful to explore briefly what is meant by war. According to Cook (2004)

What, however, is war? According to the eminent Prussian soldier and theorist Von Clausewitz (2007), "War is ...an act of violence to compel our opponents to fulfill our will." Clausewitz continues to unpack the role of violence in its relation to will. "Violence, that is to say, physical force (for there is no moral force without the conception of States and Law), is therefore the means; the compulsory submission of the enemy to our will is the ultimate object." Therefore, American warriors are, as Vice Admiral James B. Stockdale (2004) said, "in the business of breaking people's wills and the most important weapon in breaking people's wills is not the fire power but to hold the moral high ground." According to Clausewitz, Stockdale was convinced, that "War was not an activity governed by scientific laws, but a clash of wills, of moral forces." To make the same point in another way, Napoleon's succinct words are much more vivid: "In war, the moral is to the physical as three to one" (Asprey, 2002).

The direct impact of war/conflict on health is set forth by the WHO and shown in Table 3.1.

Table 3.1: The Direct Impact of Conflict on Health	
WHO Report on Violence and Health (Panch, 2004)	
Increased mortality	• Physical trauma • Infectious diseases • Deaths avoidable through health care
Increased morbidity	• Injuries due to physical trauma • Injuries due to increased societal violence • Infectious diseases • Reproductive health issues • Nutrition • Mental health
Increased disability	• Physical • Psychological • Social

3.1   IMPACT OF THE WAR IN IRAQ

An analysis of a nationwide survey of Iraqi households estimates that in excess of 100,000 deaths occurred since the 2003 invasion and possibly many more (Panch, 2004). Most deaths were the result of air strikes by coalition forces. More than half of those reported killed by coalition forces were women and children. In other words, more than 500,000 women and children have been killed in Iraq since the outset of the 2003 war. No figures are available on civilians injured during the conflict. Typically, it is estimated that the number of people injured is three times the number of deaths though evidence suggests that due to widespread terrorist activities and coalition force responses, the ratio may be as high as ten to one. That equates to likely numbers of civilian injuries ranging from 300,000 to 1,000,000 with more than half likely women and children. Compounding the tragedy of increased civilian injuries is the estimated 10 million landmines and explosive remnants of war in northern Iraq alone which could take up to approximately 15 years to clear (Pacific Disaster, 2004).

The impact of the conflict on both the health system and the health sustaining infrastructure has also been profound. With respect to various health services, there has been a marked reduction in security, financial exclusion and geographical exclusion. In Iraq, there is a shift in emphasis from primary care to specialist care, a reduction in rural and community based services and compromised public health programs. The country has seen widespread destruction of health clinics and is characterized by a lack of needed drugs and a lack of proper equipment maintenance. Transport of vaccines has been particularly difficult because of the need for sustained, cold temperatures in route.

3.2   LANDMINES

One of the most controversial weapon systems used in modern warfare is the land mine. There are between 70 and 80 million land mines in the ground in one-third of the world's nations. Landmines are indiscriminate weapons that maim or kill 15,000 to 20,000 civilians every year. They cost as little as $3 to produce, but as much as $1000 to remove.

The presence of land mines threatens people's lives, and also prevents much-needed economic growth and development. Long after wars are over, land mines make land unusable for farming, schools or living, preventing people from rebuilding lives torn apart by conflict.

Land mines are explosive material contained in casings of metal, plastic or wood that detonate from the pressure of a footstep (anti-personnel mine) or a passing vehicle (anti-tank mine). Children are just as likely to step on a land mine as a soldier. A land mine typically includes the following components: firing mechanism or other device (including anti-handling devices); detonator or igniter (sets off the booster charge); booster charge (may be attached to the fuse, or the igniter, or be part of the main charge); main charge (in a container, usually forms the body of the mine); and casing (contains all of the above parts).

Land mines are an ancient invention dating back 600 years. Precursors of conventional landmines appeared in the 15th century at the Battle of Agincourt in England and in the 18th century during the American Civil War. After his troops encountered these devices, the commander of the Union Army, General William T. Sherman, said that the use of landmines "was not war, but murder" (Sherman, 2000).

In the 20th century, land mines were developed to meet new threats including tanks and armored vehicles. More than 300 million anti-tank mines were used during World War II alone. WWII also saw an increase in the use of anti-personnel land mines. Since anti-tank mines could be removed by the enemy, anti-personnel mines were placed around them as guards. One of the most effective anti-personnel types was the German-made "bouncing betty," designed to jump from the ground to hip height when activated, propelling hundreds of steel fragments over a wide range. Noting their effectiveness, militaries began to use anti-personnel mines as weapons in their own right.

Since 1945, land mines have been used throughout the world in wars of liberation, civil wars, and local conflicts, with a devastating impact on economic and political reconstruction. Land mines demoralize survivors as well as their families and communities, impeding the process of peace and reconciliation. Often carefully and cleverly hidden, population movement is restricted, land cannot be cultivated, roads and bridges cannot be rebuilt and refugees cannot return to their homes. Survivors of land mine accidents often cannot work at their previous jobs and require retraining. Without a strong workforce, the pace of reconstruction slows.

For poor countries, clearing land mines places an additional financial resource burden. Typically, it costs from $300 to $1,000 to locate and destroy a single landmine and an additional $100 to $3,000 to provide an artificial limb to a land mine survivor. This cost does not go away as children and adults must replace these devices on a yearly basis.

The international community is making a concerted effort to eradicate land mines with the passage of the Mine Ban Treaty (also known as the Ottawa Convention) which came into force on March 1, 1999. International, nongovernmental and private-sector organizations are working with affected countries to establish mine action campaigns. As of 2008, the United States has refused to sign the treaty. According to Bush Administration, the United States will not join the Ottawa Convention because its terms would have required it to give up a needed military capability (CNN, 2004). This reverses a position previously held by the Clinton Administration (Landmines, 2008).

3.3 CONCLUSIONS AND REFLECTIONS

One of the most active individuals in mounting a public campaign to outlaw the use of land mines world wide is Bianca Jagger[1]. She has dedicated a large portion of her professional life to the issues of landmines and cluster bombs. In a recent article (Jagger, 2007), she described her position concerning the position and policy of the British government:

> "Faced with a growing body of data and expert testimony that these self-destruct mech-anisms (of the landmines placed in a war time situation) don't work, UK officials have

---

[1]Bianca Jagger is a social and human rights advocate and a former actress. Jagger is a Council of Europe Goodwill Ambassador, Chair of the World Future Council, Chair of the Bianca Jagger Human Rights Coalition, and a member of the Director's Leadership Council of Amnesty International U.S.

Figure 3.3: Global use of antipersonnel mines since May 2005. (Landmine Monitor Report 2006, International Campaign to Ban Landmines. Reposted with permission,
`http://www.icbl.org/lm/2006/`)

focused on discrediting the evidence not on examining the implications of this for the protection of civilians. Despite the rhetoric, close scrutiny shows that protection of civilians is rarely at the forefront of Government thinking on this issue. In March this year the Government announced that it had completed a review that had 'considered carefully the humanitarian factors' associated with cluster bombs. Yet officials have refused to reveal what evidence regarding civilian harm was actually considered - it is known that the assessment contained no statistics regarding civilian casualties, nor on the quantities of unexploded bombs left in different post-conflict environments, nor statistics on the areas of land painstakingly cleared by civilian teams and funded out of precious development monies."

Jagger added:

"It seems to be based on two premises - one which is misguided and the other which is simply callous. The first is a lazy assumption that protecting civilians means putting British troops at risk. Yet serious military thinkers recognize that this is not the case: in

Figure 3.4: 1997 Convention on the prohibition of the use, stockpiling, production, and transfer of antipersonnel mines and their destruction (Landmine Monitor Report 2006, International Campaign to Ban Landmines. Reposted with permission,
`http://www.icbl.org/lm/2006/`)

modern war, killing civilians fuels the enemy. The second premise is that foreign civilians do not really count. Whilst robust methods are used to protect domestic populations against anything from dangerous pharmaceuticals to faulty toys, there is no equivalent diligence when the likely victims are foreign, disempowered and voiceless."

It is the last comment that warrants further reflection. Killing civilians adds to the intensity of wars and resoluteness of the population to continue the struggle no matter the devastation. This was certainly the case in World War II for both the Germans and the Japanese. Perhaps even more disturbing is the observation that while we spend great sums and invest considerable time and effort in careful consideration of the safety of toys for our children and drugs for our population, there is virtually no consideration given to victims who may be foreign, or disempowered and or voiceless.

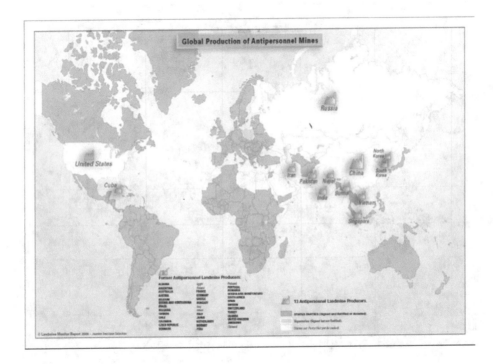

Figure 3.5: Global production of antipersonnel mines (Landmine Monitor Report 2006, International Campaign to Ban Landmines. Reposted with permission, http://www.icbl.org/lm/2006/)

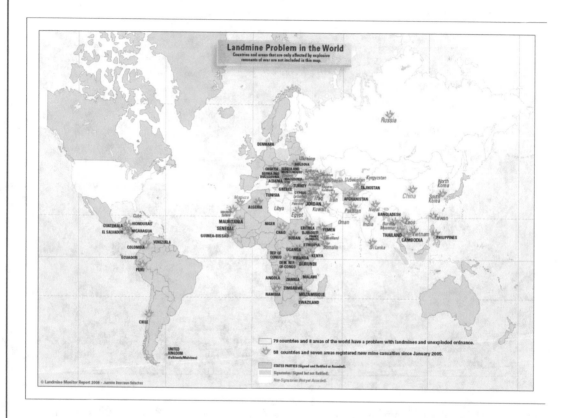

Figure 3.6: Landmine problem in the world (Landmine Monitor Report 2006, International Campaign to Ban Landmines. Reposted with permission,
http://www.icbl.org/lm/2006/)

Challenge Box: What do you personally know about the impact of war? Is your information from first hand sources or from the media? Can war ever be justified, that is, what is a just war?

Many engineers work in war-related industries. We have designed the weapon systems. Universities seek out defense related funding aggressively to support both graduate and undergraduate programs. Do we in engineering have any special sense of ethical responsibility for war? And what of the engineering curriculum, do you feel that adequate attention is paid to the issues related to conflict and war in your studies? Do you believe the issues should even be included or are those more appropriate for other fields of study?

Consider the following scenario: You finish up your formal training and get a spectacular offer from an engineering firm that consults world-wide in a wide array of topics. The firm is just awarded a new contract to design the next generation of land mines for use in protecting the interests of the U.S. abroad. Because of your expertise, you have been chosen project director...a guaranteed path to advancement. What are you thoughts and feelings concerning the possible position?

C H A P T E R   4

Hurricane Katrina and the Flooding of New Orleans

Randy Newman, *Louisiana 1927* (Common Dream).

What has happened down here is the wind have changed

Clouds roll in from the north and it started to rain

Rained real hard and rained for a real long time

Six feet of water in the streets of Evangeline

The river rose all day

The river rose all night

Some people got lost in the flood

Some people got away all right

The river have busted through clear down to Plaquemines

Six feet of water in the streets of Evangeline

CHORUS

Louisiana, Louisiana

They're tryin' to wash us away

They're tryin' to wash us away

Louisiana, Louisiana

They're tryin' to wash us away

They're tryin' to wash us away

Led Zeppelin, *When the Levee Breaks* (News Hounds, 2005).

If it keeps on rainin' levee's goin' to break

If it keeps on rainin' levee's goin' to break

When The Levee Breaks I'll have no place to stay.

Mean old levee taught me to weep and moan

Lord mean old levee taught me to weep and moan

Got what it takes to make a mountain man leave his home

Oh well oh well oh well.

Don't it make you feel bad

When you're tryin' to find your way home

You don't know which way to go?

If you're goin' down South

They go no work to do,

If you don't know about Chicago.

Cryin' won't help you, prayin' won't do you no good,

Now, cryin' won't help you, prayin' won't do you no good,

When the levee breaks, mama, you got to move.

All last night sat on the levee and moaned,

All last night sat on the levee and moaned,

Thinkin' 'bout me baby and my happy home.

Going, go'n' to Chicago,

Go'n' to Chicago,

Sorry but I can't take you.

Going down, going down now, going down.

Over the course of the last 100 years, two floods have devastated the city of New Orleans and surrounding areas: the Great Flood of 1927 and the aftermath of Hurricane Katrina in 2005. Both events have been immortalized by artists with two examples in music provided. There are many others.

4.1   THE GREAT FLOOD OF 1927

As the 1927 flood approached New Orleans, about 30 tons of dynamite were set off on the levee at Cameron, Louisiana and sent 7,000 m³/s of water pouring through. This prevented New Orleans from experiencing serious damage, but flooded much of St. Bernard Parish and all of Plaquemines Parish's east bank. As it turned out, the destruction of the Caernarvon levee was unnecessary; several major levee breaks well upstream of New Orleans, including one the day after the demolitions, made it impossible for flood waters to seriously threaten the city. There is some belief that the purpose of the levee explosion was to save the wealthier parts of the city by directing the flow of water to the more rural, less developed communities in order to minimize financial losses.

LIBRARY OF CONGRESS

Figure 4.1: Breach of levee, 1927.

U.S. ARMY CORPS OF ENGINEERS

Figure 4.2: Flooding in Morgan City, 1927.

NATIONAL ARCHIVES

Figure 4.3: Barge, near Lake Pontchatrain, 1927.

4.2   HURRICANE KATRINA

Seventy five years later an even more devastating flood would create great tragedy again for the residents of New Orleans. On August 29th, 2005, the nation watched, in horror, video images of New Orleans sliding into utter chaos. Initially, there was some hope that New Orleans had been spared the full effect of the long-predicted Category 4 hurricane[1], Katrina. This initial hope was soon replaced by horror and disbelief as two levees designed and built to hold back the waters of Lake Pontchatrain failed and the city of New Orleans began to fill with water.

Figure 4.4: NOAA satellite image of hurricane Katrina (NOAA, 2005).

The extent of the tragedy remains indelibly engraved in the minds of anyone who remembers those images. Water filled in places to the rooftops, almost immediately, in the Lower Ninth Ward, home to some of the poorest people in the area, St. Bernard Parish and New Orleans East. People, who had not been able to follow, believe or even hear the evacuation orders, were trapped in their

---

[1] Hurricane strengths are judged according to the Saffir-Simpson Hurricane Scale is a scale classifying most Western hemisphere tropical cyclones that exceed the intensities of tropical depressions and storms, and thereby become hurricanes. The categories into which it divides hurricanes are distinguished by the intensities of their respective sustained winds. The classifications are intended primarily for use in measuring the potential damage and flooding a hurricane will cause upon landfall.

homes, forced to climb to the second floors and then to the attics, finally to break though their own roofs with nothing more than a pocket knife where they waved desperately for someone to rescue them. Others waited on sidewalks, luggage packed, for the buses they thought were coming to take them to safety, eventually giving up and returning to their houses where they were trapped by the rising water.

In the aftermath of Hurricane Katrina, engineers investigated the possibility that a failure in the design, construction, or maintenance caused much of the flooding. Originally, it was speculated that the levees had been overtopped by the storm-surge; however, this was later found not to be the case. Some investigations pointed to the possibility of a weakening of the soil beneath the foundations of the flood walls due to storm water caused the ground to shift, which would indicate that a major design flaw made during the construction of the levees had been a major cause of the failures due to the storm.

Furthermore, the region's natural defenses, the surrounding wetlands and barrier islands have been dwindling in recent years. Much of the land was undeveloped wetlands on the lake side, and only small levees were constructed in the 19th century. A much larger project to build up levees along the lake and extend the shoreline out by dredging began in 1927. As the city grew, there was increased pressure to urbanize lower areas, and, as a result, a large system of canals and pumps was constructed to drain the city. Drainage of the formerly swampy ground allowed more room for the city to expand, but also resulted in subsidence of the local soil.

Outside of the city, the Mississippi River's natural deposition of suspended sediment built up the river's delta marshlands during periodic flooding episodes. However, the lower Mississippi was later restricted to channels for the benefit of shipping, which interrupted the process that continued to build the delta region south and east from New Orleans and prevented its erosion. As the swampy lands of Southern Louisiana shrank, the land began to sink. Entire barrier islands disappeared during periodic storms as the land of the vast delta slowly settled without river silt to replenish the wetlands. Approximately one-third of the land subsidence has been attributed to the large number of canals through the delta. Barge traffic and tides erode the earth around the edge of the canals, and salty water from the Gulf of Mexico water seeps in along them, slowly salinating the ground and killing the vegetation that the land previously depended on to anchor it.

The flooding of metro New Orleans was an avoidable man-made disaster. The levee and canal walls failed because of human errors. "Experts say the New Orleans flood of 2005 should join the space shuttle explosions and the sinking of the Titanic on history's list of ill-fated disasters attributable to human mistakes" (Gelinas, 2007).

The U.S. Army Corps of Engineers, a federal agency has sole authority over the design and construction of metro New Orleans' flood protection and water management as authorized by Congress in the Lake Pontchartrain Hurricane Protection Project. The U.S. Army Corps of Engineers has accepted that faulty design specifications and substandard construction of certain levee segments, not a hurricane were the primary cause of the flooding damage in the New Orleans area.

Figure 4.5: Aerial view of the flooding in part of the central business district (The Superdome is at center.) (Access Kansas, 2005).

Figure 4.6: Flooded I-10 interchange and surrounding area of northwest New Orleans and Metairie, Louisiana (Wikipedia, 2008).

Today, almost three years have passed. Throughout much of New Orleans, recovery from Katrina has been hindered by the city's many pre-storm weaknesses and delayed by false starts. Entire city blocks of the Lower 9th Ward closest to the Industrial Canal breach are now a field of prairie grass pockmarked with concrete-slab foundations and driveways ending in wildflowers. Community anchors are boarded up or abandoned, scarred by fire and rotting from water. The federally funded Road Home program, the largest housing recovery program in U.S. history, has given out $6.4 billion to storm victims looking to sell or rebuild their homes. There seems to be a growing momentum for restoration as well as something much more important, hope.

> **Challenge Box:** Should we care about New Orleans? Why or why not? As engineers, do we have a special responsibility for the Katrina-caused damage to the city and the loss of life? Should we help New Orleans rebuild? Should the residents of the city be allowed to return?

Unfortunately, there's no escaping the blocks of warped and ruined homes. Pontchartrain Park, north of Gentilly and bordering the lake, was occupied primarily by retirees. Most residents who lived near Lake Pontchatrain were overwhelmed by the tragedy and the needed effort to come back, and as a result, areas such as Pontchatrian Park and Gentilly remain ghost towns. The city-wide population of New Orleans is far below its pre-Katrina level of 454,000. The numbers are in dispute, most reliable estimates are around 308,000—a rebound of about 180,000 since the months after the storm. Most agree resettlement is leveling off. New homes are being built higher and "smarter," but the emphasis for flood protection remains with the levees. While the Army Corps of Engineers argues that dramatic repairs have significantly strengthened the system, critics like Robert Bea, a civil engineering professor at the University of California-Berkeley, are dismissive. "It may be able to withstand a Category 3 storm," he says, "but not likely." That level of protection won't be fully in place until 2011. And it's unlikely the system will ever be strong enough to protect against a Category 5 storm."

> **Challenge Box:** The levee system was an engineering failure, yet all specifications/codes were met. What does that say to you about the notion of safety? Is there a responsibility beyond the regulations, the codes and the laws?
> Many of the citizens of New Orleans who could not escape the city before the storm hit were the poor or elderly, children or the infirmed. Is there any additional responsibility for these individuals? Why or why not?

More than 150,000 jobs were lost in the storm. The city has had difficulty even ascertaining who owns what property, especially in the Lower Ninth Ward, where many houses passed through generations of families with no changes in the deeds. And at least half the city's pre- storm residents were renters, few of them with insurance. Now the city is aggressively pushing home ownership for all.

CHAPTER 5

Disappearing Bumble Bees

We should preserve every scrap of biodiversity as priceless while we learn to use it and come to understand what it means to humanity.

E.O. Wilson *(Brainy Quote, 2008).*

One way to think about the sun, every time you see it at dawn, is to think of it as an act of cosmic generosity.

Brian Swimme *(Wikipedia, 2008).*

Figure 5.1: E.O. Wilson.

Nature is a mutable cloud which is always and never the same.

Ralph Waldo Emerson *(Quoteworld, 2008).*

Figure 5.2: Emerson.

In the U.S., bees are responsible for pollinating up to one-third of the food supply, yet they have been dying off in alarming numbers for almost two decades. Faced with such detrimental foes as parasitic mites, pesticides, and urban development, bees are at a formidable disadvantage in our modern world. But since crops are so dependent on the work of these tiny creatures, beekeeping has become a big business, and maintaining effective hives is practically a competitive sport. Honeybees, even as they dwindle in number, are the most dependable source of pollination, carrying much of the burden to sustain our agricultural needs.

Figure 5.3: Honey bees serve as pollinators.

5.1   DEVELOPING CRISIS

Honey bees pollinate more than 90 cultivated crops, including avocadoes, cucumbers, watermelons, citrus fruit and, notably, almonds; California's almond industry alone needs about half the country's 2.5 million commercial hives for pollination every year. Honey bees are responsible for more than $20 billion in annual pollination value and one-third of the food we eat, from vegetables to oils to meat, from animals that graze on pollinated foliage. Managed honey bees are not the only pollinators in the air as, in the Pacific Northwest alone, there are some 900 species of native bees. Some plants, such as tomatoes, rely on wind pollination. But because modern agriculture demands high yields from densely planted crops, they need modern commercial honey bees in massive quantities.

Wild bees and the flowers they pollinate are disappearing together in Britain and the Netherlands, researchers have announced. It is not clear which started to disappear first, the bees or the flowers, but the trend could affect both crops and wild species, the researchers report in a recent issue of the journal ScienceDaily (2007).

An undiagnosed honeybee ailment spread across the Northern Hemisphere has left British beekeepers worrying about how many of their bees survived the winter. The implications for agricul-

Figure 5.4: Honey bees in marked and mysterious decline.

tural pollination and production are huge. No one knows the cause of the honeybee sickness, which caused the deaths of thousands of honeybee colonies across the US and Europe.

This follows a series of unexplained, but very severe, honeybee colony losses over the past few years in Poland, Greece, Italy, Spain and Portugal and heavy losses in other countries are suspected to be going unreported. The symptoms of the colony deaths are varying across Europe and North America and the losses generally come to light between late summer and early spring. This winter in the USA, colonies have dwindled as the older bees have died leaving behind the queen and young workers not yet ready to forage for pollen and nectar and insufficient in number to maintain the colony. In the United Kingdom this past year, there were a few but significant examples of what became termed the Marie Celeste phenomenon - colonies simply disappearing from hives leaving no bees for post-mortem analysis.

Challenge Box: Should we care about rapid decline of bumble bees? Why or why not? Do bees matter as pollinators for our food or do they matter in their own right? Do we need to worry more generally about the loss of wildlife? What can we do about this decline? Is there a connection between the plight of the bumble bees and engineering? If it exists, what is that connection?

In the closing months of 2006, thousands of American bee hives were found to be almost entirely devoid of bees, victims of a mysterious phenomenon now known as Colony Collapse Disorder (CCD). A study of 150,000 managed bee colonies in 15 states, commissioned by the Apiary

Inspectors of America, found that from September 2006 to March 2007, roughly one-third of the colonies were lost.

Bee keepers have suffered similar unexplained losses in the past, and not all of the hives in the survey were lost to whatever is causing colony collapse. But people are understandably worried that the disorder may threaten all three million managed bee colonies in the United States, a $14.6 billion commercial pollination business. As a result, it has become urgent that scientists determine what is causing the colonies to disappear and how many more colonies stand to vanish.

5.2   COLONY COLLAPSE DISORDER MODEL

Many scientists have suggested that some kind of virus or bacterium — or some combination of infectious agents, possibly carried by parasites like mites — is killing the bees. A model was developed and examined in 2003 to determine whether severe acute respiratory syndrome could be controlled (Ellis, 2008). A similar model may be used to study CCD.

A colony which has collapsed from CCD is generally characterized by all of these conditions occurring simultaneously: (a) a complete absence of adult bees in colonies, with little or no build-up of dead bees in or around the colonies; (b) a presence of capped brood in colonies. Bees normally will not abandon a hive until the capped brood have all hatched; (c) Presence of food stores, both honey and bee pollen: which are not immediately robbed by other bees and which when attacked by hive pests, the attack is noticeably delayed. Precursor symptoms that may arise before the final colony collapse are: (a) an insufficient workforce to maintain the brood that is present; (b) a workforce made up of young adult bees; (c) a queen is present; and (d) colony members are reluctant to consume provided feed.

Challenge Box: How do you react to the assertion that bio-engineering is either a futile exercise or an ecological disaster with no possibility for recall? Explain your position.
Explain in your own words what is meant by a mechanical universe mindset. Do you have one? Give examples which reinforce your assertion.

Perhaps the disappearance of the bee colonies point to far broader issues that need to be considered such as the diminishing biodiversity in our planet's ecosystem, the widespread use of pesticides and herbicides, unintended gene transfer from genetically modified crops, direct and indirect stress on the ecosystem, global climate change, navigational hindrances and, in our opinion the most significant, a "mechanical universe" mindset. Such a mindset reduces an infinitely complicated world of interactions to an overly simplistic viewpoint. In mathematics 1 + 1 = 2, in biology, 1 + 1 may equal 3, or a billion and three. Some argue that the term bio-engineering itself is a contradiction in terms. 'Bio' equates to 'life.' 'Engineering,' as we have seen earlier in this text, refers to something entirely different. Biological forms can never be 'engineered' - i.e., predictably controlled or manipulated. Unlike a sheet of metal that can be machined with consistent results, organisms in natural systems are ever changing and adjusting. This makes 'bio-engineering,' in the best-case scenario, a futile exercise and an enormous misallocation of human and environmental resources, and, in the worse case scenario, an ecological catastrophe with no chance for a product recall.

CHAPTER 6

Engineering and Traditional Approaches

A man is truly ethical only when he obeys the compulsion to help all life which he is able to assist, and shrinks from injuring anything that lives.

Albert Schweitzer.

Figure 6.1: Albert Schweitzer.

As stated described, we shall begin our exploration of making decisions using approaches that are very commonly utilized in engineering today: utilitarianism, rights of persons and virtue ethics. An *Utilitarian* approach emphasizes bringing about the most good that we can. A *rights of person* approach states that actions or decisions are right if each person or "moral agent" is afforded equal respect and consideration. An approach based on *virtue* is identified as the one that emphasizes the virtues, or moral character, in contrast to the approach which emphasizes duties or rules or that which emphasizes the consequences of actions.

6.1 TRADITIONAL APPROACHES TO ENGINEERING ETHICS

6.1.1 UTILITARIANISM

According to Hinman (2008), the basic insights offered by Utilitarianism can be listed as follows: the purpose of morality is to make the world a better place; morality is about producing good consequences not having good intentions; and we should always choose based on bringing the

How is one to live a moral and compassionate existence when one is fully aware of the blood, the horror inherent in life, when one finds darkness not only in one's culture but within oneself? If there is a stage at which an individual life becomes truly adult, it must be when one grasps the irony in its unfolding and accepts responsibility for a life lived in the midst of such paradox. One must live in the middle of contradiction, because if all contradiction were eliminated at once life would collapse. There are simply no answers to some of the great pressing questions. You continue to live them out, making your life a worthy expression of leaning into the light.

Barry Lopez.

Figure 6.2: Barry Lopez.

greatest good to the greatest number of people. Harris et al. (2008) offer the following standard for utilitarianism: those individual actions or rules that produce the greatest total amount of utility to those affected are right. Here the most common definition of utility is happiness. utilitarians are in agreement that for most people to be able to pursue utility or happiness, two conditions must be present: freedom and well-being. Freedom is defined as the power to exercise choice and make decisions without constraint from within or without; autonomy; self-determination while well-being amounts to how well a person's life is going for him or her. It may include health, material well-being, food, shelter and education or training. Three approaches to Utilitarianism are the cost/benefit approach, the act utilitarian approach and the rule utilitarian approach.

Harris et al. (2008) suggest the following three steps for implementing a cost/benefit analysis approach to making engineering decisions:

- Assess the available options

- Assess in monetary terms the costs and benefits for all of those affected

- Make the decision that is most likely to result in the greatest benefit for the least cost

Cost/benefit analysis (CBA) is a generic term for a variety of techniques designed to allow decision-makers to determine in a rigorous way whether the payback from a program will be greater than the costs of implementing it. For example, if costs of an environmental program are greater than environmental benefits produced by a program, the program should be abandoned. The economic justification for this use of CBA is the notion that society must decide how to spend its scarce resources and it should spend its money in the most efficient way possible. If money is spent by society on environmental protection programs that do not produce an environmental payback that is greater in economic value than the cost of the program, it is a bad investment and should not be supported. According to CBA theory, public money should be spent on programs that will produce the largest aggregate benefits.

Act Utilitarianism is based on the notion that it is the value of the consequences of *the particular act* that counts when determining whether the act is right. Act Utilitarianism makes no appeals to general rules, but instead demands that the person(s) making the decision evaluate individual circumstances. Harris et al. (2008) offer the following procedure for implementation of an Act Utilitarian approach:

- Determine the possible option available

- Determine the appropriate audience for each of the options

- Remember the principle of *universalizability* must be met and remain consistent. (Immanuel Kant used this term when discussing the maxims, or subjective rules, that guide our actions. A maxim is *universalizable* if it can consistently be willed as a law that everyone ought to obey. The only maxims which are morally good are those which can be universalized. The test of universalizability ensures that everyone has the same moral obligations in morally similar situations

- Select the option that maximizes the good for the previously determined relevant audience

The third utilitarian approach is referred to as Rule Utilitarianism. "Instead of looking at the consequences of *a particular act*, Rule Utilitarianism determines the rightness of an act by a different method. First, the best rule of conduct is found. This is done by finding the value of the consequences of *following a particular rule*. The rule the following of which has the best overall consequences is the best rule. Among early proponents were Austin (1995) and Mills (2008). Rule Utilitarianism is an option for those who believe that there are absolute prohibitions on certain types of actions but do not want to give up on utilitarianism completely. According to Rule Utilitarianism, the principle of utility is a guide for choosing rules, not individual acts.

Challenge Box: Compare and contrast a cost benefit analysis approach, an Act Utilitarian approach an a Rule Utilitarian approach to the following case:

A trolley is running out of control down a track. In its path are 5 people who have been tied to the track by a mad philosopher. Fortunately, you can flip a switch which will lead the trolley down a different track to safety. Unfortunately, there is a single person tied to that track. Should you flip the switch?

As before, a trolley is hurtling down a track towards five people. You are on a bridge under which it will pass, and you can stop it by dropping a heavy weight in front of it. As it happens, there is a very fat man next to you - your only way to stop the trolley is to push him over the bridge and onto the track, killing him to save five. Should you proceed?

6.1.2   RESPECT FOR PERSONS

According to Harris et al. (2008), the moral standard of the ethics of *respect for persons* is straightforward: "Those actions or rules are right that, if followed, would accord equal respect to each person as a moral agent." A moral agent is a person with a capacity for making moral judgments and taking actions that comport with morality. A moral agent is a being capable of those actions that have a moral quality, and which can properly be denominated good or evil in a moral sense. Most philosophers tend to view morality as a transaction among rational parties, i.e., among moral agents, and thus, would exclude animals as well as land and ecosystems from moral consideration. Others (Singer, 2005) state that one must draw a distinction between moral agency and being subject to moral considerations and have argued that the key to inclusion in the moral community is not rationality — for if it were, we might have to exclude some disabled people and infants, and might also have to distinguish between the degrees of rationality of healthy adults — but that the real object of moral action is the avoidance of suffering.

One *respect for persons* approach with which we are very familiar is the Golden Rule, a variant of which appears in the religious and ethical writings of societies and cultures throughout the world (Table 6.1). The Golden Rule is best interpreted as saying: "Treat others only in ways that you're willing to be treated in the same exact situation." To apply it, you are asked to imagine yourself in the exact place of the other person on the receiving end of the action. If you act in a given way toward another, and yet are unwilling to be treated that way in the same circumstances, then you violate the rule.

To apply the Golden Rule appropriately, we need knowledge and imagination. We need to *know* what effect our actions have on the lives of others. In addition, we need to be able to *imagine* ourselves, vividly and accurately, in the other person's place on the receiving end of the action. Application of the Golden Rule can take the following steps:

- Determine the effects on others lives that will result from our action(s).

- Imagine ourselves in the other persons place, that is, 'walk a mile in that person's shoes.'

- Ask ourselves, would we be willing to accept the consequences of our action(s)?

Table 6.1: Comparison of Golden Rule for various cultures and religions.

Culture/Religion	Statement of the Golden Rule
Buddhism	*Hurt not others in ways that you yourself would find hurtful.* Udana-Varga 5,1
Christianity	*All things whatsoever ye would that men should do to you, do ye so to them; for this is the law and the prophets.* Matthew 7:1
Confucianism	*Do not do to others what you would not like yourself. Then there will be no resentment against you, either in the family or in the state.* Analects 12:2
Hinduism	*This is the sum of duty; do naught onto others what you would not have them do unto you.* Mahabharata 5,1517
Islam	*No one of you is a believer until he desires for his brother that which he desires for himself.* Sunnah
Jewish	*What is hateful to you, do not do to your fellowman. This is the entire Law; all the rest is commentary.* Talmud, Shabbat 3id
Taoism	*Regard your neighbor's gain as your gain, and your neighbor's loss as your own loss.* Tai Shang Kan Yin P'ien
Zoroastrianism	*That nature alone is good which refrains from doing another whatsoever is not good for itself.* Dadisten-I-dinik, 94,5

A second *respect for persons* approach is referred to as the Rights Based Approach. According to the Stanford Encyclopedia of Philosophy (2008). "Rights are entitlements (not) to perform certain actions or be in certain states, or entitlements that others (not) perform certain actions or be in certain states. Rights dominate most modern understandings of what actions are proper and which institutions are just. Rights structure the forms of our governments, the contents of our laws, and the shape of morality as we perceive it. To accept a set of rights is to approve a distribution of freedom and authority, and so to endorse a certain view of what may, must, and must not be done." Other philosophers and ethicists suggest that the ethical action is the one that best protects and respects the moral rights of those affected. This approach starts from the belief that humans have a dignity based on their human nature per se or on their ability to choose freely what they do with their lives. On the basis of such dignity, they have a right to be treated as ends and not merely as means to other ends. The list of moral rights -including the rights to make one's own choices about what kind of life to lead, to be told the truth, not to be injured, to a degree of privacy, and so on-is widely debated; some now argue that non-humans have rights, too. Rights imply duties, in particular, the duty to respect others' rights.

Challenge Box: Consider the same two scenarios set forth in the trolley problem but this time applying the Golden Rule first and then the Rights Based approach.

6.1.3   VIRTUE ETHICS

According to its etymology the word virtue (Latin *virtus*) can be interpreted to mean 'possessing courage.' Virtues may be divided into intellectual, moral, and theological. As we are focusing upon engineering and making decisions, we shall limit our discussion to moral virtues only. One particular list of moral virtues includes the following elements:

- Justice: Justice regulates man in relations with his fellow-men. It disposes us to respect the rights of others, to give each man his due. Justice, a condition there of, is the ideal state of humanity, a morally correct state of things and persons. John Rawls[1] claims "Justice is the first virtue of community institution, as truth is of systems of thought." Today many people believe that justice must not be limited or restrained. Many social and political movements are centered, in fact, on the premise of global justice.

- Temperance: Temperance represents a certain quality of self-control and discipline. Individual power lies in slowly tempering desires so that they disappear. Justice then lies in tempering the balance of power so that all maintain equal shares.

---

[1]John Rawls (1921-2002) is considered by many to be one of the most important political philosophers of the 20th century. He wrote a series of highly influential articles in the 1950s and '60s that helped refocus Anglo-American moral and political philosophy on substantive problems about what we ought to do.

Figure 6.3: Virtue.

- Fortitude: Fortitude, which implies a certain moral strength and courage, is the virtue by which one meets and sustains dangers and difficulties, even death itself. It refers to that strength or firmness of mind which enables a person to encounter danger with coolness and courage, or to bear pain or adversity without complaint.

The term "virtue ethics" is a relatively recent one. It is an umbrella term that encompasses a number of different theories. Initially, virtue ethics was characterized as a movement rivaling consequentialism and deontology because it focused on the central role of concepts like character and virtue in moral philosophy. Later versions developed fuller accounts of virtue ethics theories. Most virtue ethics theories take their inspiration from Aristotle, although some versions incorporate elements from Plato[2], Aquinas,[3] Hume[4] and Nietzsche.[5]

[2] Plato (c. 427-347 B.C.E) is one of the world's best known and most widely read and studied philosophers. Known as the student of Socrates and the teacher of Aristotle, he wrote in the middle of the fourth century B.C.E.

[3] Saint Thomas Aquinas, (1225–1274) was a Dominican priest, a philosopher and a theologian in the scholastic tradition. He was the foremost classical proponent of natural theology, and the originator of the Thomistic school of philosophy and theology.

Virtue ethics is concerned with the person's life as a whole, with character and the kind of person you are. The right perspective on an action, therefore, will for virtue ethics be the one which asks about success in achieving the overall goal, rather than success in achieving the immediate target. What matters is what the person's motivation was, and how this relates to his or her developed character and life as a whole; for this is his or her achievement, what he or she has made of his or her life. Success or failure in achieving the immediate target will affect various judgments we make about the action, but if, like the Stoics, we distinguish clearly between the immediate target and the overall aim, it is achieving the latter, not the former, which will make the action a success. Here virtue ethics differs from theories like consequentialism, for which it is the actual results that matter for our evaluation of the agent (Annas, 2008), and agrees with Kantianism (Kant, 1780), for which what matters is the agent's motivation.

Challenge Box: Let's return to the same trolley problem scenarios. Compare and contrast your responses when you apply a virtue ethics based approach versus the other applied ethical approaches you have already employed. What is the same? What is different?

Which ethical approach was the easiest for you to use? Which one proved most troublesome? Explain.

Did any of the approaches yield answers which you found contrary to your own personal values/beliefs? Explain. Did you find that troubling?

6.2   REVIEW OF EXISTING ETHICAL CODES

At the start of the 21st century, there are as many different codes of conduct in engineering as there are engineering disciplines and specialties. One professional society, the National Society for Professional Engineers (NSPE), has offered one general code which is widely employed today in all the disciplines as well as in engineering education. The NSPE Code of Ethics consists of a preamble followed by a listing of fundamental canons and then rules of practice (NSPE, 2008). The very first canon cautions engineers in the fulfillment of their professional duties, to "hold paramount the safety, health and welfare of the public." As a result, the first rule of practice states that engineers shall "hold paramount the safety, health, and welfare of the public." Note that the explicit requirements focus

---

[4] David Hume (1711-1776) is considered by many to be the most important philosopher ever to write in English. Today, philosophers recognize Hume's work as a precursor of contemporary cognitive science, as well as one of the most thoroughgoing exponents of philosophical naturalism.

[5] Frederick Nietzche (1844-1900) was a German scholar, philosopher and critic of culture.

on the public only; though presumably, concern for the natural world is included implicitly, though only, as it affects humankind.

The American Society of Mechanical Engineers (ASME) sets forth a similarly constructed code of ethics with fundamental principles followed by fundamental canons (ASME, 2008). The first principle states that engineers uphold and advance the integrity, honor, and dignity of the engineering profession by using their knowledge and skill for the enhancement of human welfare. The supportive fundamental canon states engineers shall hold paramount the safety, health and welfare of the public in the performance of their professional duties.

The American Society of Civil Engineers (ASCE) does, at least, mention the environment in its code (ASCE, 2008). According to ASCE, engineers uphold and advance the integrity, honor and dignity of the engineering profession by using their knowledge and skill for the enhancement of human welfare and the environment (fundamental principle) and shall hold paramount the safety, health and welfare of the public and shall strive to comply with the principles of sustainable development in the performance of their professional duties (fundamental canon). There is no explanation of what is meant by the enhancement of the environment. In November 1996, the ASCE Board of Direction adopted the following definition of sustainable development: "Sustainable development is the challenge of meeting human needs for natural resources, industrial products, energy, food, transportation, shelter, and effective waste management while conserving and protecting environmental quality and the natural resource base essential for future development."

The Institute of Electrical and Electronics Engineers (IEEE) Code of Ethics states that its members accept responsibility in making engineering decisions consistent with the safety, health and welfare of the public, and to disclose promptly factors that might endanger the public or the environment (IEEE, 2008). Here, an interesting notion of responsibility towards the environment is described. It is not in opposition to the IEEE code to endanger the public or the environment only to not disclose promptly factors that might endanger the public or the environment.

The Institute of Industrial Engineers (IIE) endorses the Canon of Ethics provided by the Accreditation Board for Engineering and Technology (ABET) whose first principle is that engineers uphold and advance the integrity, honor and dignity of the engineering profession by using their knowledge and skill for the enhancement of human welfare and whose first canon is engineers shall hold paramount the safety, health and welfare of the public in the performance of their professional duties (IIE, 2008). ABET is the accrediting body for all engineering and engineering technology programs in the United States and thus has an important impact on the training of tomorrow's engineers and engineering educators.

Members of the American Institute of Chemical Engineers (AIChE) are challenged to uphold and advance the integrity, honor and dignity of the engineering profession by being honest and impartial and serving with fidelity their employers, their clients, and the public; striving to increase the competence and prestige of the engineering profession; and using their knowledge and skill for the enhancement of human welfare (AIChE, 2008). To achieve these goals, AIChE members shall hold paramount the safety, health and welfare of the public and protect the environment in performance of their professional duties. There is neither elaboration on the idea of protecting the environment nor an identification on from whom or what shall it be protected.

Many other engineering disciplines exist, each with their own codes for ethical conduct. As can be seen from this review, a large percentage of the codes do not explicitly identify the environment as an important stakeholder in discussions of the ethics of engineering choices. Equally as troubling, those codes that do mention the environment refer to the idea of enhancing nature or promoting sustainable development, which is based solely upon meeting human needs. A select few number of codes do mention a responsibility to protect the environment but without identifying from whom or from what.

There are many other engineering disciplines at present, each with its own code of conduct or ethics, which describes the responsibilities of the profession. Most focus heavily on the sense of responsibility engineering has towards employers, society in general and towards other professional engineers.

In their totality, the codes of ethics point to a very different conception or understanding of the natural world then our science provides us with now. We are at once removed from membership in the natural world as there is a listing of responsibilities of the engineering profession to humankind, and if it exists at all, a sense of responsibility to the natural world only in so far as it can provide something for us. We are not products of the earth but somehow placed on it with a focused plan of action set in place to tame it, control it, and to transform it into what suits are interests.

Challenge Box: Compare and contrast the various codes of ethics. Which provided insights when confronted with issues of war, poverty and the environment? Which seemed to closest match your own set of values? Which do you think would be easiest to follow? Which one do you think would be most troublesome? Explain your responses.

Did any of the approaches yield answers which you found contrary to your own personal values/beliefs? Explain. Did you find that troubling?

CHAPTER 7

Engineering and Freedom

Figure 7.1: Freedom.

To this point in our explorations of approaches to making decisions in the 21st century, we have examined traditional applied ethics methodologies; many of which date back to antiquity. We shall now explore several modern notions which may aid us as engineers as we confront ever more difficult choices. The first approach we shall examine is an outgrowth of the ethics of freedom.

7.1 INDIVIDUALISM AND FREEDOM

In the West, the modern conception of humankind is characterized, more than anything else, by individualism. The implications of this individualism give rise to our understanding of freedom. One

> *No man can put a chain about the ankle of his fellow man without at last finding the other end fastened about his own neck.*
>
> *Frederick Douglass[a].*

[a]Frederick Douglas - speech, Civil Rights Mass Meeting, Washington, D.C., 1883.

Figure 7.2: Douglass.

effort to fully explore the notion of freedom can be found in existentialist theory[1]. An existentialist conception of individuality may give rise to the following set of questions relevant for our search for approaches to confronting serious questions in the 21st century:

- What is human freedom?

- What can the absolute freedom of absolute individuals mean?

- What is human flourishing or human happiness?

- What general ethic or way of life emerges when we take our individuality seriously?

- What ought we to do?

- What ethics or code of action can emerge from a position that takes our individuality seriously?

Sartre[2] explores an appropriate ethics code using existentialist theory. According to Sartre, we each individually choose human nature for all humans. Hence, we must choose courses of action that we would wish all humans to take. In choosing for ourselves, we choose for all of humanity. Thus, I must choose in the same way we would want others to choose - another instance of the use of the Golden Rule. We speak of acting authentically when we ignore the external differences among ourselves and other people as these differences are merely outward manifestations of who we are —

---

[1]Existentialism is a philosophy that emphasizes the uniqueness and isolation of the individual experience in a hostile or indifferent universe, regards human existence as unexplainable, and stresses freedom of choice and responsibility for the consequences of one's acts.

[2]Jean Paul Sartre (1905-1980) was a French novelist, playwright, existentialist philosopher, and literary critic. Sartre was awarded the Nobel Prize for literature in 1964, but he declined the honor in protest of the values of bourgeois society.

not the essence of who we are. Sartre also argues that in order to be free, we must desire the freedom of all humanity. It is self-defeating to attempt to use other humans as objects to satisfy our desires, or to protect our freedom at the cost of enslaving others. The person who uses other people as objects to satisfy his desires makes himself or herself an object. To see others as slaves of our desire is to make ourselves a slave of desire. Thirdly, our decisions are not arbitrary as we speak of a coherence of our actions. Our actions must unify the many different influences on our lives into the one life that is to be ours. Our actions, though free, are constrained by our situation in a community with all its relationships and obligations.

In summary, freedom as constructed from an existentialist perspective, must take on the responsibility of choosing for all of humankind, desire and work for the freedom of all humanity, and create ourselves within the context of the relationships and obligations we have to others.

7.2   ETHICS OF FREEDOM: SUBSTANTIAL FREEDOM

Sen[3] is seen as a ground-breaker among late 20th century economists for his insistence on discussing issues seen as marginal by most. He mounted one of the few major challenges to the economic model (capitalism) that has placed self-interest as the prime motivating factor of human activity. Sen (Amartya, 2005) describes what he refers to as the basic idea of positive or substantial freedom, distinguishes it from negative freedom on the one side and happiness on the other, and relates it to the notion of "capability," which is distinct from a raw capacity and an actual exercise of a capability. He further states that "the perspective of freedom" is concerned with "enhancing the lives we lead and the freedoms we enjoy." The ethic of freedom calls for, "expanding the freedoms we have reason to value," so that our lives will be "richer and more unfettered" and we will be able to become "fuller social persons, exercising our own volitions and interacting with–and influencing–the world in which we live."

Garret (2008) offers the following clarification of what Sen means by 'substantial freedom.'

According to Garret, "Substantial freedoms are valuable things that can be divided up and delivered to human beings (or groups of people in a region) in varying amounts. In that respect they are like money and freedom from coercion, things which can be preconditions for substantial freedom but are not very good indicators of it. In the case of money, a person can have very little, a middle amount, or a lot. In the case of freedom from coercion, the same can be said: one can be a slave, constantly subject to the whims of an overseer, or one can have maximum available freedom from coercion by one's fellow humans in their private or governmental capacities. What society does, and to some extent what individuals do, can determine how much substantial freedom we have."

Continuing, "substantial freedom is distinguished from other things that we 'often have reason to value:' money, negative freedom or freedom from coercion, and happiness, on the other. Monetary income alone cannot be used as a reliable indicator of substantial freedom. An increase in income

---

[3] Amartya Kumar Sen is an Indian economist, philosopher, and a winner of the Nobel Prize for Economics in 1998, "for his contributions to welfare economics" for his work on famine, human development theory, welfare economics, the underlying mechanisms of poverty, and political liberalism.

might be converted into an increase in substantial freedom, but the conversion is not automatic or equally easy for everybody. A sick person is normally less able than a healthy one to convert a given increase in income into a wider range of real opportunities, i.e., into greater substantial freedom. The same might be said of a person who lives in a dangerous neighborhood that makes him/her fearful to go outside as compared to a person who lives in a safer neighborhood" (Garret, 2008).

7.3   ETHICS OF FREEDOM: CAPABILITIES

Nussbaum (1999) has continued to develop the notion of substantial freedom. According to Nussbaum, "At the heart of this tradition is a twofold intuition about human beings: namely, that all, just by being human, are of equal dignity and worth, no matter where they are situated in society, and that the primary source of this worth is a power of moral choice within them, a power that consists in the ability to plan a life in accordance with one's own evaluation of ends." To these two ideas is linked one more, that "the moral equality of persons gives them a fair claim to certain types of treatment at the hands of society and politics. This treatment must do two things: respect and promote the liberty of choice, and …respect and promote the equal worth of persons as choosers."

A necessary component of Nussbaum's approach is the list of basic capabilities. She answers the question, "What activities characteristically performed by human beings are so central that they seem definitive of a life that is truly human?" Nussbaum's list includes the following:

- The ability to live life to its natural end

- Maintaining health and integrity of the body

- The ability to move freely about and be free from the threat of violence

- Being able to use the senses; being able to imagine, to think, and to reason

- Being able to have attachments to things and persons outside ourselves; being able to love those who love and care for us while not having one's emotional developing blighted by fear or anxiety.

- Being able to form a conception of the good and to engage in critical reflection about the planning of one's own life.

- Being able to live for and in relation to others, to recognize and show concern for other human beings, to engage in various forms of social interaction; being able to imagine the situation of another and to have compassion for that situation; having the capability for both justice and friendship.

- Being able to be treated as a dignified being whose worth is equal to that of others.

- Being able to live with concern for and in relation to animals, plants, and the world of nature.

- Being able to laugh, to play, to enjoy recreational activities.

- Possessing control over one's environment.

Challenge Box: Suppose you were challenged to put an engineering ethic based on freedom in your own words and have it be less than fifty words. What would you write? What similarities would there be between this approach versus the other more traditional approaches we have already taken? What differences?

Let's go back to our trusted trolley problem. What would an ethic based on freedom point to for both scenarios? Explain. Are the recommendations different than what you concluded earlier?

C H A P T E R   8

Engineering and Chaos

Figure 8.1: Chaos.

In the twentieth century, science is undergoing a major reshuffling. The work of Einstein and others has shown that Newtonian science adequately describes a limited number of idealized problems at best. A new science, the science of chaos, is beginning to supplant the classical mechanics of Newton. It is a science of disorder, probabilities and non-linearities. Most interestingly, it appears to be a more accurate description of nature, for nature's essence is chaos. The winds of the atmosphere, the streams of the oceans, the sliding of the surface platelets all are chaotic. Chaos also describes the deposition of river silt along the Mississippi Delta, the flow of water through the giant oak trees,

> *I accept chaos. I am not sure whether it accepts me. I know some people are terrified of the bomb. But then some people are terrified to be seen carrying a modern screen magazine. Experience teaches us that silence terrifies people the most.*
>
> *Bob Dylan.*

Figure 8.2: Dylan.

> *You need chaos in your soul to give birth to a dancing star.*
>
> *Nietzsche.*

Figure 8.3: Nietzsche.

and the flow of blood in the arteries and veins of human beings. The nonlinear interaction of one human-being with another is also chaotic.

8.1   DEFINITION

What is chaos? In fact, many pundits or 'talking-heads' that appear regularly in the media claim our world is becoming ever more chaotic so its probably useful for us to understand what is meant by the term. The notion of chaos is linked to change which we can categorize into a few fundamental categories: growth and recession, stagnation, cyclic behavior and unpredictable, erratic fluctuations. Nearly all of these phenomena can be described with well developed *linear* mathematical tools. Here *linear* means that the result of an action is always proportional to its cause: if we double our effort, the outcome will also double. Most of nature is non-linear in the same sense as most of zoology is non-elephant zoology (Ulam, 0000). Scientists who study chaos and its broader are of specialty, non-liner mechanics, often liken the situation that most of traditional science is focusing on linear systems can be compared to the story of the person who looks for the lost car keys under a street lamp because it is too dark to see anything at the place where the keys were lost. With recent advances in computational speed, memory and computing algorithms, scientists now have the ability to make significant progress in the field of non-linear systems and begin to understand phenomena that we had no hope of understanding in the past.

This phenomenon, *chaos*, exists outside the framework of linear theory. The modern notion of *chaos* describes irregular and highly complex structures in time and in space that follow deterministic laws and equations which is in contrast to the totally random chaos of classical thermodynamics.

Mayer-Kress offers the following visualization (Figure 8.4) to illustrate the difference among determinism, chaos and randomness (Mayer-Kress, 1995).

"Consider a fluid in a pot on a stove in which the level of stress is given by the rate at which the fluid is heated. At room temperature, the water is in equilibrium with its surroundings is totally placid. The appearance of placidity indicates a total lack of spatial or temporal coherence. As heat is added to the water, the system starts to self-organize and form regular spatial patterns (rolls, hexagons) which create coherent behavior of the subsystems ('order parameters slave subsystems'). The order parameters themselves do not evolve in time. Under increasing stress the order parameters themselves begin to oscillate in an organized manner: we have coherent and ordered dynamics of the subsystems. Further increase of the external stress leads to bifurcations to more complicated temporal behavior, but the system as such is still acting coherently. This continues until the system shows temporal deterministic chaos. The dynamics is now predictable only for a finite time. This predictability time depends on the degree of chaos present in the system. It will decrease as the system becomes more chaotic. The spatial coherence of the system will be destroyed and independent subsystems will emerge which will interact and create temporary coherent structures."

This flow field is termed a Benard flow field in which convection cells that appear spontaneously in a liquid layer when heat is applied from below. In this configuration, we have a thin layer of water or some other liquid filling the gap between two parallel plates. The bottom plate is then heated. A simple sketch of the resulting flow field is shown in Figure 8.5.

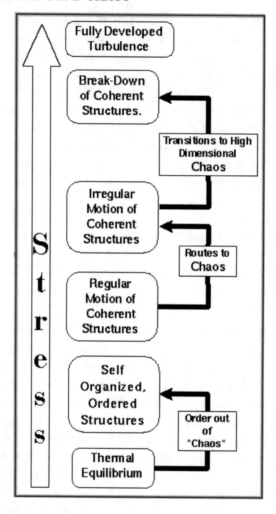

Figure 8.4: Onset of turbulence: chaos to order & order to chaos (Mayer-Kress, 1995).

8.2   NATURE AND CHAOS

Our efforts to understand the shift in perspective offered by chaos may be best illustrated by first considering the implications chaos has for our understanding of the natural world. After exploring these implications, we will ask the following questions: How does chaos affect the way I may approach decision-making in the 21st century? What new insights does it offer?

Nature still was thought to be "naturally" in an equilibrium condition. An example of the Newtonian view of the stability and orderliness of the natural world is to be found in the writings of George Perkins Marsh (Marsh, 2008).

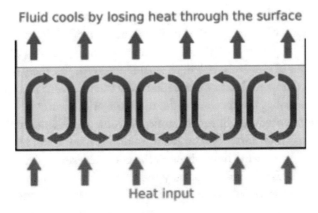

Figure 8.5: Schematic of a Benard Cell.

Nature, left undisturbed so fashions her territory as to give it almost unchanging permanence, outline, and proportion, except when shattered by geologic convulsions; and in these comparatively rare cases of derangement, she sets herself at once to repair the superficial damage, and to restore, as nearly as practicable, the former aspect of her dominion.

Newtonian mechanics views the universe as a gigantic mechanism that functions like clockwork, and represents the ultimate in reliability and mechanical perfection. Newtonian mechanics casts the laws of nature in the form of mathematical equations (Stewart, 1998). The key point about these equations is that each of their solutions is unique. For example, a bicycle has perhaps some four or five moving parts. If the motion of each of the parts one instant of time is known, then Newtonian mechanics allows the determination of the motion of the bicycle at some later point farther down the road. Cosmically, the knowledge of the position and the velocities of every point of matter in the Solar System at one instant in time would allow the unique determination of all subsequent motions of those particles at some later time. This statement both assumes for mathematical simplicity that there are no external influences on this motion, and leads to the interpretation that the positions and the velocities of every particle in the entire universe taken at some fixed instant of time completely determines its future evolution. The entire universe follows a unique, predetermined path according to the French mathematician Laplace [from Stewart (1998)]:

"An intellect which at any given moment knew all the forces that animate Nature and the mutual position of the beings that comprise it, if this intellect were vast enough to submit its data to analysis, could condense into a single formula the movement of the greatest bodies of the universe and that of the lightest atom: for such an intellect nothing would be uncertain; and the future just like the past would be present before its eyes."

This statement by Laplace describes the paradigm of classical determinism. The paradigm worked well in the solutions of many technical problems and provided the mathematical foundation for the West's "Industrial Revolution." There was great optimism, for it seemed only a matter of time until all the problems of classical mechanics would be solved. Yet, there remained one problem in Newtonian mechanics: how to describe mathematically the turbulent motion of a fluid.

Turbulent flow describes the motion of the natural world. Nearly all motion that occurs in nature is turbulent. Turbulence is nonlinear, dynamic, random, multi-dimensional, and is characterized by indeterminacy and intermittency. Turbulent motion is the motion of the clouds, the earth's winds and seas, and the flow of living fluids through the major arteries of living creatures (plants and animals alike). Turbulence also describes not only the interaction of one member of a species with another, but also the interaction of one species with another. Prior to the latter half of the twentieth century, scientists believed that the differential equations, independently developed by Navier and Stokes, would effectively predict the laminar flow of a fluid and would also prove adequate for the challenge of turbulent flow. They anticipated that the advent of faster and larger computers or ever more esoteric and *ad hoc* models would provide the missing link in the search for the deterministic solution.

Such hopes appear today to be naive at best or arrogant at worst, for a deterministic solution for the turbulent flow of a fluid is no nearer at hand than it was at the turn of the 20th century. Rather than the old paradigm of classical determinism, a paradigm based on chaos more clearly describes turbulence. While Newtonian mechanics enables the determination of exact solutions to a narrow set of idealized problems, it has not been adequate in scientists' attempts to model and predict turbulent fluid flow. Rather than being an uncommon occurrence, the turbulent motion of a fluid is the motion of life and reflects the essence of the natural world. The new science of chaos holds potential for a clearer understanding of turbulence and, hence, a clearer understanding of the natural world.

In a fluid, we have turbulent cascades where vortices are created that will decay into smaller and smaller vortices. Analog situations in societies can be currently studied in the former USSR and Eastern Europe. James Marti speculates: "Chaos might be the new world order" (Mayer-Kress, 1995). At the limit of extremely high stress, we are back to an irregular chaos where each of the subsystems can be described as random and incoherent components without stable, coherent structures. It has some similarities to the anarchy with which we started close to thermal equilibrium. Thus, the notion of *chaos* covers the range from completely coherent, slightly unpredictable, strongly confined, small scale motion to highly unpredictable, spatially incoherent motion of individual subsystem.

8.3   NEWTONIAN AND CHAOS-BASED ETHIC

In *Sand County Almanac* (Leopold, 1968), Leopold wrote:

> "All ethics so far evolved rest upon a single premise: that the individual is a member of a community of interdependent parts. Ecology simply enlarges the boundaries to include soils, waters, plants, and animals, or collectively the land. In short a land ethic changes

the role of Homo *Sapiens* from conqueror of the land community to plain member and citizen of it. It implies respect for his fellow members and also respect for the community as such."

Leopold went on to formulate "The Land Ethic:"

A thing is right when it tends to preserve the integrity, stability, and beauty of the biotic community. It is wrong when it tends otherwise.

Leopold's view of the natural world is apparent when he asserts that the *stability of Nature is disrupted by the interference of Man.* He believed that Nature left to its own design will "naturally" opt for the stable or equilibrium position and that it is Man who disrupts this stability, challenges its integrity, and causes a natural world with diminished beauty. This view is a clear precursor of the thought of Thomas Berry, who compared the beauty of the Hudson River before its occupation by the Western world to its present condition and found its grandeur diminished.

The failure of the mechanical model of the natural world can be demonstrated through various examples[1]. Sometimes more than simply being folly, the Newtonian view of nature led to ecological disasters.

If we shift our model of the natural world from its historic deterministic base to one which embraces chaos, a new ethic can be suggested:

A thing is right when it tends to allow the natural world and all the entities thereof, to thrive in richness and diversity, and to experience change. It is wrong when it tends otherwise.

Note that there are three elements to this new ethic.

- Richness here refers to the richness of experience of the various entities that make up the natural world (Johnson, 1991).

- Diversity refers to wide variety in plant and animal species. An action which would result in the enhancement of the variety of species that would exist in a given ecosystem would be in accord with the new environmental ethic.

- Change as the restraint of change is a violation of the processes that model the natural world.

---

[1] Probably no other controversy has done more to divide the ranks of conservationists around the world or more to cripple ecological research in East Africa than that involving the elephants of Tsavo. Because of pressures outside the huge 8,300 square-mile Tsavo National Park, thousands of elephants have been crowding into the sanctuary, swelling the population to somewhere between 20,000 and 30,000, and making it the largest remaining concentration of elephants in the world. Because elephants eat so much and push down so many trees in the course of their activities, large concentrations can devastate an area. Tsavo, once a lush land of dense bush and trees is today a sad landscape of bare ground and struggling grasses, strewn with the skeletons of downed trees. Because the park receives very little rain, some scientists have said the destruction could eventually turn Tsavo into a desert, providing so little food that most of the elephants would eventually starve to death. The long-term result could be a "population crash" that would wipe out elephants in the very place set aside to protect them. Three years ago a drought reduced the already damaged vegetation so severely that between 5,000 and 6,000 elephants died of malnutrition. Since then, still more elephants have entered the park, seeking refuge from hunters and expanding agriculture. The population may again be as high as before the drought. Further droughts and massive die-offs are almost certain in coming months and years. When a scientific team recommended shooting 3,000 elephants for research purposes and indicated it might be necessary to shoot many more to bring the population back into balance with its environment, many conservationists recoiled in horror. Killing the animals they were trying to protect hardly fit the traditional conservation ethic.

Challenge Box: Compare and contrast a vision of a mechanical universe versus one that is based on chaos. Which one makes the most sense according to your values/beliefs?
What similarities would there be between this chaos based approach versus the ethic of freedom and other more traditional approaches we have already taken? What differences? Let's go back to our trusted trolley problem. What would an ethic based on chaos point to for both scenarios? Explain. Are the recommendations different than what you concluded earlier?

C H A P T E R   9

Engineering and a Morally Deep World

> *Geese appear high over us, pass, and the sky closes. Abandon, as in love or sleep, holds them to their way, clear in the ancient faith: what we need is here. And we pray, not for new earth or heaven, but to be quiet in heart, and in eye, clear. What we need is here.*
>
> *Wendell Berry.*

Figure 9.1: Berry.

A new code of ethics is offered for engineering adapted from an environmental model of nature as a self-organizing system. A self-organizing system is characterized by synthesis rather than analysis and suggests a new code of ethical responsibility based upon community rather than individuality.

9.1   SELF-ORGANIZING SYSTEMS

Modern science at the start of the 21st century does not model the natural world using either the great chain of being or the mechanical clock paradigms or as living being (Gaia hypothesis). Today, the natural world is most often described using the model of a self-organizing system and nature rather than being thought of as immutable is seen as constantly in change. Self-organization refers to a process in which the internal organization of a system, normally an open system automatically without being guided or managed by an outside source. Self-organizing systems typically (though not always) display emergent properties. Emergence is the process of complex pattern formation from simpler rules. This can be dynamic (occurring over time), such as the evolution of the human brain over thousands of successive generations; or emergence can happen over disparate length scales, such as the interactions between a macroscopic number of neurons producing a human brain capable of thought (even though the constituent neurons are not themselves conscious). For a phenomenon to be termed emergent, it should generally be unpredictable from a lower level description.

The world abounds with systems and organisms that maintain a high internal energy and organization in seeming defiance of the laws of physics. According to Decker (1995), "As a bar of iron cools, ferromagnetic particles magnetically align themselves with their neighbors until the entire bar is highly organized. Water particles suspended in air form clouds. An ant grows from a single-celled zygote into a complex multicellular organism, and then participates in a structured hive society. What is so fascinating is that the organization seems to emerge spontaneously from disordered conditions, and it does not appear to be driven solely by known physical laws. Somehow, the order arises from the multitude of interactions among the simple parts. The laws that may govern this self-organizing behavior are not well understood, if they exist at all. It is clear, though, that the process is nonlinear, using positive and negative feedback loops among components at the lowest level of the system, and between them and the structures that form at higher levels."

Decker adds:

"The study of landscape ecology provides an example of how an SOS perspective differs from standard approaches. Ecologists are interested in how spatial and temporal patterns such as patches, boundaries, cycles, and succession arise in complex, heterogeneous communities. Early models of pattern formation use a 'top-down' approach, meaning the parameters describe the higher hierarchical levels of the system. For instance, individual trees are not described explicitly, but patches of trees are. Or predators are modeled as a homogenous population that uniformly impacts a homogeneous prey population. In this way, the population dynamics are defined at the higher level of the population, rather than being the results of activity at the lower level of the individual."

Finally,

"The problem with this top-down approach is that it violates two basic features of biological (and many physiochemical) phenomena: individuality and locality. By modeling a rodent population as a mass of rodents with some growth and behavior parameters, we obviate any differences that might exist between individual rodents. Some are big, some are small, some reproduce more, and some get eaten more. These small differences can lead to larger differences - such as changes in the population gene frequencies, individual body size, or population densities - that might have cascading effects at still higher levels. The tenet of locality means that every event or interaction has some location and some range of effect. This is a simple illustration of the ecological principle that pattern affects process. To say that a system is self-organized is to say it is not entirely directed by top-down rules, although there might be global constraints on the system. Instead, the local actions and interactions of individuals generate ordered structures at higher levels with recognizable dynamics. Since the origins of order in SOS are the subtle differences among components and the interactions among them, system dynamics cannot usually be understood by decomposing the system into its constituent parts. Thus, the study of SOS is synthetic rather than analytic."

If the self-organized system is used to model the natural world rather than the great chain of being or the mechanical clock, our sense of responsibilities to the natural world seem to change significantly. We are forced to look "synthetically" rather than "locally," that is at the very least or moral sphere of concern must broaden. Secondly, nature is no longer in perfect order nor is it a collection of parts, i.e., gears, levers, weight) which can be replaced or modified according to our desires. The mechanical clock in many ways has been replaced by a seemingly chaotic clock which defies predictability, single-valueness and repeatability. If we are to make sense of our place in this natural world, we need a very different sense of ethics. One attempt at providing such an ethical framework has been offered by Johnson (1993) in his development of a morally deep world.

9.2   A MORALLY DEEP WORLD

According to Leopold, acting ethically is a matter of concern both for us and for others with whom we are in some sort of community. The notion of a community deserves some discussion. We, perhaps, are most comfortable with community referring to a body of people having common rights, privileges, or interests, or living in the same place under the same laws and regulations; as, a community of Franciscan monks. In biology or ecology, community refers to an interacting group of various species in a common location. For example, a forest of trees and undergrowth plants, inhabited by animals and rooted in soil containing bacteria and fungi, constitutes an integral community (Johnson, 1993). Extending the notion of community in this way is consistent with the pattern evidenced in human society over the centuries. We have progressively enlarged the boundaries of our understanding of community and recognized the membership of slaves, foreigners, etc., those for whom membership was not extended at earlier times in history. Leopold's land ethic then "simply enlarges the boundaries of the community to include soils, waters, plants, and animals, or collectively: the land."

Johnson discusses how non-sentient land can count morally and focuses upon the concept of a living being (Johnson, 1993). For Johnson, a living being is best thought of not as a thing of some sort but as a living system, an ongoing life-process. A life-process has a character significantly different from those of other processes such as thermodynamics processes, for example. Our character, as living beings, is the fundamental determinant of our interests. Johnson adds further that:

"The interests of a being lie in whatever contributes to its coherent effective functioning as an on-going life-process. That which tends to the contrary is against its interests....moral consideration must be given to the interests of all living beings, in proportion to the interest. Some living systems other than individual organisms are living entities with morally considerable interests. ...All interests must be taken into account."

The concept of a morally deep world was developed within the framework of environmental ethics. Perhaps, it may be useful to explore the morally deep world argument as it applies to a specific and presently quite contentious issue in wildlife management today, the reintroduction of the Mexican wolf into regions of the Southwestern United States. For the purposes of illustration, let us focus on the land near the White Sands Missile Range near Las Cruces, New Mexico. Johnson would challenge us to first identify all the members of the community. For this example, a listing would include the following:

- Wolves

- Prey animals including domestic sheep and cattle as well as deer, rabbits, coyotes, and others

- Desert lands

- Ranchers and sheep farmers

- Hunters

- U.S. Fish and Wildlife Service and other state and local government agencies

- U.S. Department of Defense

- Residents of White Sands and nearby towns and settlements

- Residents of New Mexico and the entire United States

- Native American residents

Often in such cases, two very different perspectives dominate the deliberations. On one side of the debate is atomism, a view that moral assessment applies only to individuals. The individual would be individual wolves, prey, ranchers, etc. On the other side is holism, a view that collectives or whole are subject to moral appraisal. In a morally deep world, the view is shortsighted morally if one adopts either a holistic or atomistic. No one (holistic or atomistic) interest has priority over the other. There is an inevitable tension between atomistic and holistic ethics. Sometime the interests of the biotic community will outweigh the interests of the individual, while at other times it is the interests of the individual which are paramount. Let us next identify the extent of the community or living being in this case. Recall that a living being is characterized as having an ongoing life process with interests in whatever contributes to its coherent effective functioning. Clearly, wolves, their prey, the desert lands, ranchers, sheep farmers, hunters and people who live in or near White Sands have considerable interests. Other identified elements could be argued to have less interest in the coherent effective functioning of the community. That is not to suggest that, for example, the residents in New York would have no interest in the restoration, but their impact on the coherent effective functioning of the ongoing process would be less.

An interesting example of the tension between atomism and holism can be identified in the following scenario. Suppose wolves are restored to the White Sands Missile range desert and suppose that, as has been the case in Yellowstone National Park, wolves adapt well and quickly grow in numbers. In Yellowstone, some wolves are routinely killed as part of wolf or game management practices. From a holistic perspective this may be morally acceptable, but it would be difficult to justify the killing from an atomistic perspective. A morally deep world point of view would argue that both interests need to be considered carefully, including the interests of the entire park community and those of the "surplus" wolf.

One criticism often offered of a morally deep world perspective is that it prevents any action that will affect a community. On the contrary, though a morally deep perspective does assert actions that violate vital interests of the community or erosion of its self-identity should be avoided, it requires active participation in the protection of the essential functions and the maintenance of the viability of life processes. Rather than calling for inaction, a morally deep world perspective suggests contemplation followed by direct and specific responses.

9.3   ENGINEERING IN A MORALLY DEEP WORLD

Given a shift to a morally deep world paradigm, a new engineering code of ethics is outlined. The majority of existing codes are structured, in similar if not identical ways, with fundamental principles supported by fundamental canons. That same structure will be incorporated into the present work. For a morally deep world, the first fundamental canon and rule of practice is specified as:

Engineers, in the fulfillment of their professional duties, shall hold paramount the safety, health and welfare of the identified integral community.

The fundamental difference between an ethical code based on a morally deep world versus the present codes is the replacement of the "public" by the "identified integral community."

How would decisions be made in engineering adhering to a morally deep world perspective? The following flow chart is offered with the decision process consisting of the following four steps:

- Via Positiva. The problem is identified, fully accepted and broken down into its various components using the vast array of creative and critical thinking techniques which engineers possess. What is to be solved? For whom is it to be solved?

- Via Negativa. Reflection on the possible implications and consequences for any proposed solution are explored. What are the ethical considerations involved? The societal implications? The global consequences? The effects on the natural environment?

- Via Creativa. The third step refers to the act of creation. The solution is chosen from a host of possibilities, implemented and then evaluated as to its effectiveness in meeting the desired goals and fulfilling the specified criteria.

- Via Transfomativa. The fourth and final step asks the following questions of the engineer: Has the suffering in the world been reduced? Have the social injustices that pervade our global village been even slightly ameliorated? Has the notion of a community of interests been expanded? Is the world a kinder, gentler place borrowing from the Greek poet Aeschylus? (Smith, 1970).

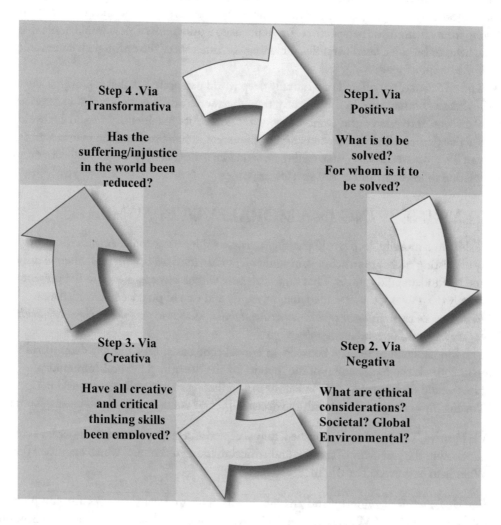

Figure 9.2: Engineering decision algorithm based upon a Morally Deep World.

9.4   APPLYING THE MORALLY DEEP WORLD VIEW

Consider the following scenario. Harvesting machines are replacing migrant workers as grape pickers in Northern California (Tesconi, 2000). Today, such machines cost approximately $126,000. It takes five workers about eight hours to pick 10 tons of grapes. To harvest the 100 tons would require 90 workers if they wanted to be done before sun-up. The costs are relatively straightforward to calculate.

It costs $120 a ton to harvest by hand, and $30 a ton by machine. A mid-size wine producer expects to harvest about 4,000 tons of grapes total. For 4,000 tons, the cost to pick the grapes by hand is approximately $480,000 while the costs using a harvester are $120,000 representing an increase in margin above cost equal to $340,000. Even with the cost of the harvester subtracted from this margin, the net increase in profits is equal to $204,000.

Traditional engineering ethics discussions might stop here. But what if we go a bit farther and identify the workers as being part of the community? What else would we be forced to consider? If it takes 90 workers to harvest 100 tons then it would take the equivalent of 3600 workers to harvest 4000 tons. Certainly there are many more people involved than simply the workers themselves with the total involved growing far beyond the estimate of 3600. What will become of them? Should we as the engineers who designed, built, tested, evaluated and delivered the harvester care? If we use the code of ethics described by countless engineering societies today, the answer would be no. There is no consideration given to the workers whose livelihoods have been eliminated. Codes of engineering ethics based on the notion of a morally deep world would suggest a very different result as it forces us as engineers to consider the entire living system, in this case, including the men, women and children who toil as grape pickers and whose quality of life is intimately linked to the harvest.

Figure 9.3: An example of a mechanical fruit picker.

For this case, the following questions concerning the significance of the grape-harvesting device may be asked:

- For whom should the device be designed? The owners? The land? The workers? Someone else?

- What will become of the displaced workers?

- What will become of the displaced workers families?

- Are there long-term effects on the land?

- Are there societal implications for the workers' communities?

- Are there societal implications for the greater community outside the workers and their families?

The identified members of the community may include:

- Owners

- Workers

- Workers' families

- Land and water ecosystem

- Farm worker society

- Local, regional and national societies

Suppose we use the proposed engineering design algorithm. In Via Positiva, an identification of the problem is made. Grapes have to be picked at a particular time and in a particularly rapid fashion. The harvesting of grapes by hand has in fact been done satisfactorily for thousands or years on farms of all sizes and now to a limited extent by machines. In Via Negativa, the possible implications or consequences of the change to a harvesting machine are considered. The effects upon the environment for the given case seem minimally affected by the mechanization of the process as stated here. The consequences for the workers, their families and their ways of life are however profound and disturbing. A large number of workers will be displaced and their jobs completely eliminated. In Via Creativa, engineers have, in fact, designed and delivered machines that will meet the criteria set forth by the grape farmers and have thus met their professional responsibilities as has been commonly understood. But a consideration of the net effects on the suffering in the world, Via Transformativa, might yield a different result. Engineers who design such devices without concern for the impact on the numbers of farm workers who are being replaced solely in order to increase the profits of the landowners may have acted unethically.

In the context of the farm workers discussed here, a logical question to ask would be the following: Is it possible to arrive at an end result which was a creative design that met everyone's needs and established justice for workers and their families and furthered the interests of the vineyard? While we cannot offer a device, we can suggest that those whom the new machine would make irrelevant might be included in discussions concerning the consequences of implementing the new design. The landowner may ultimately decide to mechanize this task but at the very least the criteria for making ethical choices would also include a careful consideration of the impact of the engineering design on the lives of the workers. It would seem such an inclusion is as important as any other.

Challenge Box: Summarize in your own words what is meant by a morally deep world. Suppose you were challenged to write down a morally deep world ethic for engineering in less than 100 words, what would you write?

What similarities would there be between this morally deep world approach versus other more traditional approaches we have already taken? What differences?

Let's go back to our trusted trolley problem. What would an ethic based on a morally deep world point to for both scenarios? Explain. Are the recommendations different than what you concluded earlier?

C H A P T E R   10

Engineering and Globalism

> *It is yet another Civilized Power, with its banner of the Prince of Peace in one hand and its loot-basket and its butcher-knife in the other.*
>
> *Mark Twain (1992).*

Figure 10.1: Mark Twain.

There is little doubt that the world is becoming more and more globalized. The current form of globalization which is based upon neo-liberalism, free trade and open markets has sparked much debate throughout the world. Some critics would argue that the interests of powerful nations and corporations are shaping the terms of world trade. Proponents would argue that globalization has led to massive increases in the world's wealth while critics would suggest that while a few people are becoming increasingly wealthy, a greater percentage of the world's population is become poorer.

10.1 GLOBALISM AND GLOBALIZATION

Globalism, at its core, seeks to describe and explain a world which is characterized by networks of connections that span multi-continental distances. It attempts to understand all the inter-connections of the modern world — and to highlight patterns that underlie and explain them. In contrast, globalization refers to the increase or decline in the degree of globalism. It focuses on the forces, the dynamism or speed of these changes. Globalism is the underlying basic network, while globalization refers to the dynamic shrinking of distance on a large scale.

Globalization in its literal sense is the process of making, transformation of some things or phenomena into global ones. It can be described as a process by which the people of the world are unified into a single society and function together. This process is a combination of economic, technological, socio-cultural and political forces. Globalization is often used to refer to economic globalization, that is, integration of national economies into the international economy through trade, foreign direct investment, capital flows, migration, and the spread of technology.

Friedman (2007) "examines the impact of the 'flattening' of the globe," and argues that "globalized trade, outsourcing, supply chaining , and political forces have changed the world permanently, for both better and worse. He also argues that the pace of globalization is quickening and will continue to have a growing impact on business organization and practice. In the book, Friedman recounts a journey to India when he realized globalization has changed core economic concepts. He suggests the world is "flat" in the sense that globalization has leveled the competitive playing fields between industrial and emerging market countries. In his opinion, this flattening is a product of a convergence of personal computer with fiber-optic micro cable with the rise of work flow software. He termed this period as Globalization 3.0, differentiating this period from the previous Globalization 1.0 (which countries and governments were the main protagonists) and the Globalization 2.0 (which multinational companies led the way in driving global integration)."

Chomsky[1] argues that the word globalization is also used, in a doctrinal sense, to describe the neo-liberal form of economic globalization. Chomsky (1995) surprisingly agrees with Adam Smith[2] that at the outset that:

> "free movement of people is a core component of free trade. As for free movement of capital, that's a totally different matter. Unlike persons of flesh and blood, capital has no rights, at least by Enlightenment/classical liberal standards. As soon as we bring up the matter of free movement of capital, we have to face the fact that while people are in principle at least equal in rights, in a just society, talk of capital conceals the reality: we are speaking of owners of capital, who are vastly unequal in power, naturally. In the real world, free movement of capital entails radical restriction of democracy, for obvious reasons that have long been well understood. Speaking of capital and labour as if they were on a par is so hopelessly misleading that sensible discussion is impossible in these terms."

He develops the term "just globalization" in which he describes globalization as he thinks it should be when issues of democratic versus authoritarian control of production, distribution, interchange, information, etc., are all carefully considered and weighed in the conversations.

Daly suggests that sometimes the terms internationalization and globalization are used interchangeably but there is a slight formal difference (Daly, 2000). The term "internationalization" refers to the importance of international trade, relations, treaties etc. International means between or among nations. "Globalization" means erasure of national boundaries for economic purposes; international trade becomes inter-regional trade.

---

[1]Noam Chomsky is an American linguist, philosopher, political activist, author and lecturer at the Massachusetts Institute of Technology.

[2]Adam Smith (1723-1790) was a Scottish moral philosopher and a pioneering political economist. Smith is known for his explanation of how rational self-interest and competition, operating in a social framework which ultimately depends on adherence to moral obligations, can lead to economic well-being and prosperity. He is widely acknowledged as the "father of economics."

There are four distinct dimensions of globalism: economic, military, environmental and social:

- Economic globalism involves long-distance flows of goods, services and capital and the information and perceptions that accompany market exchange.

- Environmental globalism refers to the long-distance transport of materials in the atmosphere or oceans or of biological substances such as pathogens or genetic materials that affect human health and well-being.

- Military globalism refers to long-distance networks in which force, and the threat or promise of force, are deployed.

- The fourth dimension is social and cultural globalism. It involves movements of ideas, information, images and of people, who carry ideas and information with them.

Discussion on globalism primarily focuses on economic globalism. This phenomenon seems particularly important in the practice of engineering and we shall explore the implications of this form and suggest an ethical framework for making decisions in light of its existence.

With respect to economic globalism, the number of people living in absolute poverty has increased from a billion five years ago to 1.2 billion today according to a collaborative report prepared by the World Bank, the International Monetary Fund, the Organization for Economic Cooperation and Development and the United Nations. For more than 30 of the poorest national economies, real per capita incomes have been falling for the past 35 years. Asia is the only region in which poverty rates decreased during the past five years. Economic progress in Latin America was made ineffective by the increase in inequality among the various classes of society. People in the industrial countries now are 74 times richer than those in the poorest. The wealth of the three richest men in the world is greater than the combined GNP of all of the least developed countries - 600 million people. This impoverishment has occurred at a time when globalization was supposed to have launched the poor into sustained economic growth.

10.2 EMERGENCE OF A GLOBAL ETHIC

On September 4, 1993, for the first time in the history of religion, delegates to the Parliament of the World's Religions in Chicago adopted a "Declaration toward a Global Ethic"[3]. On September 1, 1997, again for the first time, the InterAction Council of former heads of state or government

---

[3]Declaration towards a Global Ethic, Chicago, 1993. The world is in agony. The agony is so pervasive and urgent that we are compelled to name its manifestations so that the depth of this pain may be made clear. Peace eludes us – the planet is being destroyed – neighbors live in fear – women and men are estranged from each other – children die! This is abhorrent. We condemn the abuses of Earth's ecosystems. We condemn the poverty that stifles life's potential; the hunger that weakens the human body; the economic disparities that threaten so many families with ruin. We condemn the social disarray of the nations; the disregard for justice which pushes citizens to the margin; the anarchy overtaking our communities; and the insane death of children from violence. In particular, we condemn aggression and hatred in the name of religion. But this agony need not be. It need not be because the basis for an ethic already exists. This ethic offers the possibility of a better individual and global order, and leads individuals away from despair and societies away from chaos. We are women and men who have embraced the precepts and practices of the world's religions: We affirm that a common set of core values is found in the teachings of the religions, and that

called for a global ethic and submitted to the United Nations a proposed "Universal Declaration of Human Responsibilities," designed to underpin, reinforce and supplement human rights from an ethical angle. In addition, the third Parliament of the World's Religions, held in Cape Town in December 1999 issued "A Call to Our Guiding Institutions," based on the Chicago Declaration.

What is, in fact, meant by a *global ethic?* By Global Ethic is meant the necessary minimum of common values, standards and basic attitudes or alternatively a minimal basic consensus relating to binding values, irrevocable standards and moral attitudes, which can be affirmed by all religions despite their undeniable dogmatic or theological differences and also be supported by non-believers. In many ways, it is a call for a change of consciousness. In its widest sense it includes all our sensations, thoughts, feelings, and volitions–in fact, the sum total of our mental life.

Challenge Box: Summarize in your own words what is meant by global ethic. Suppose you were challenged to write down a global ethic for engineering in less than 100 words, what would you write?

What similarities would there be between this global ethic approach versus other more traditional approaches we have already taken? What differences?

Let's go back to our trusted trolley problem. What would a global ethic point of view point to for both scenarios? Explain. Are the recommendations different than what you concluded earlier?

these form the basis of a global ethic. We affirm that this truth is already known, but yet to be lived in heart and action. We affirm that there is an irrevocable, unconditional norm for all areas of life, for families and communities, for races, nations, and religions. There already exist ancient guidelines for human behavior which are found in the teachings of the religions of the world and which are the condition for a sustainable world order. We declare: We are interdependent. Each of us depends on the well-being of the whole, and so we have respect for the community of living beings, for people, animals, and plants, and for the preservation of Earth, the air, water and soil. We take individual responsibility for all we do. All our decisions, actions, and failures to act have consequences. We must treat others as we wish others to treat us. We make a commitment to respect life and dignity, individuality and diversity, so that every person is treated humanely, without exception. We must have patience and acceptance. We must be able to forgive, learning from the past but never allowing ourselves to be enslaved by memories of hate. Opening our hearts to one another, we must sink our narrow differences for the cause of the world community, practicing a culture of solidarity and relatedness. We consider humankind our family. We must strive to be kind and generous. We must not live for ourselves alone, but should also serve others, never forgetting the children, the aged, the poor, the suffering, the disabled, the refugees, and the lonely. No person should ever be considered or treated as a second-class citizen, or be exploited in any way whatsoever. There should be equal partnership between men and women. We must not commit any kind of sexual immorality. We must put behind us all forms of domination or abuse. We commit ourselves to a culture of non-violence, respect, justice, and peace. We shall not oppress, injure, torture, or kill other human beings, forsaking violence as a means of settling differences. We must strive for a just social and economic order, in which everyone has an equal chance to reach full potential as a human being. We must speak and act truthfully and with compassion, dealing fairly with all, and avoiding prejudice and hatred. We must not steal. We must move beyond the dominance of greed for power, prestige, money, and consumption to make a just and peaceful world. Earth cannot be changed for the better unless the consciousness of individuals is changed first. We pledge to increase our awareness by disciplining our minds, by meditation, by prayer, or by positive thinking. Without risk and a readiness to sacrifice there can be no fundamental change in our situation. Therefore, we commit ourselves to this global ethic, to understanding one another, and to socially beneficial, peace-fostering, and nature-friendly ways of life. We invite all people, whether religious or not, to do the same.

CHAPTER 11

Engineering and Love

Figure 11.1: Love.

Love and compassion are necessities, not luxuries.
Without them, humanity cannot survive.

If the love within your mind is lost and you see other beings as enemies, then no matter how much knowledge or education or material comfort you have, only suffering and confusion will ensue.

Dalai Lama (1998).

> *Morals were too essential to the happiness of man, to be risked on the uncertain combinations of the head. [Nature] laid their foundation, therefore, in sentiment, not in science.*
>
> *Thomas Jefferson.*

Figure 11.2: Jefferson.

Engineering based on love? We have encountered the more traditional applied ethics, an ethic based on freedom, another on chaos and still others based on a morally deep world view or globalism. Each offers us different ways to respond when confronted with ethical dilemmas. Our last approach to confronting difficult decisions will be an engineering ethic based on love. Key elements of an engineering ethic based upon love would include the capacity for true, rigorous critical thought, the development of a culture in which individual dissent is honored and revered and in which each of us considers our self a citizen of the Earth. Lastly, an engineering ethic would enable each of us to develop our own individual narrative of moral imagination, that is, to develop the ability to be in another's shoes, to cultivate our inner eye of seeing and knowing and to overcome the blindness that we have all become far too accustomed. First, we shall explore the definitions of love used in this exploration.

11.1   DEFINITIONS OF LOVE

Countless sages, scholars, poets, philosophers, theologians and others have tried to define love throughout the ages. We shall use the following description of three aspects of love which may impact the manner in which we reach decisions.

- *Agape*: love that promotes overall well-being when confronted by that which generates ill-feeling (i.e., returning good for ill)

- *Eros*: love that promotes overall well-being by affirming the valuable or beautiful

- *Philia*: love that promotes overall well-being when cooperating with others.

11.2   AN ETHIC OF LOVE

Putting the three ideas together, we have the basic framework for reaching ethical decisions. Our work then must promote the overall well-being of all, including our perceptions of friends, and foes. It challenges us to reflect on the words found in nearly all universal wisdom traditions. One example is given below:

> *But I say to you that hear, love your enemies, do good to those who hate you, bless those who curse you, pray for those who abuse you. To those who strike you on the cheek offer the other also, and from those who take away your cloak, do not withhold your coat as well. Give to everyone who begs from you, and of those who take away your goods, do not ask them again. And as you wish that others would to do you, do so to them* (Luke 6, ESV Bible, 2008).

It calls for moving beyond that false dualism, false as who is to say what actually is good and what is ill? Can we ever really know with certainty? Making decisions, using a sports metaphor, ought not to be equivalent to cheering for one side over another whether it is American football or that variety of football they play in every other corner of the world.

Secondly, how can we affirm the beautiful and the valuable? Perhaps the first step is to examine that which we view as beautiful and that which we view as valuable. The subjective experience of "beauty" often involves the interpretation of some entity as being in balance and harmony with nature which may lead to feelings of attraction and emotional well-being. In its most profound sense, beauty may engender a significant or important experience of positive reflection about the meaning of one's own existence. An "object of beauty" is anything that reveals or resonates with personal meaning. Inner beauty is a concept used to describe the positive aspects of something that is not physically observable. Qualities including kindness, sensitivity, tenderness or compassion; creativity and intelligence have been said to be desirable since antiquity. In turn, let us consider what each of these qualities is asking from us.

- Kindness: the act or the state of charitable behavior to other people.

  > *"To give pleasure to a single heart by a single kind act is better than a thousand head-bowings in prayer."*
  > *Saadi*

- Sensitivity: the quality or condition of being sensitive, that is, the capacity to respond to stimulation.

  > *"Either I'm too sensitive, or else I'm gettin' soft,"*
  > *Bob Dylan*

- Tenderness: the quality or state of being considerate or protective.

  > *"Courage is by no means incompatible with tenderness. On the contrary, gentleness and tenderness have been found to characterize the men, no less than the women, who have*

done the most courageous deeds."
Samuel Smiles

- Compassion: the human feeling of pity over another's sorrows, along with the desire to help others in their situations.

  "A human being is a part of the whole called by us universe, a part limited in time and space. He experiences himself, his thoughts and feeling as something separated from the rest, a kind of optical delusion of his consciousness. This delusion is a kind of prison for us, restricting us to our personal desires and to affection for a few persons nearest to us. Our task must be to free ourselves from this prison by widening our circle of compassion to embrace all living creatures and the whole of nature in its beauty."
  Albert Einstein

- Creativity: the ability to see something in a new way, to see and solve problems no one else may know exists, and to engage in mental and physical experiences that are new, unique, or different.

  "The world is but a canvas to the imagination."
  Henry David Thoreau

- Intelligence: a property of mind that encompasses many related mental abilities, such as the capacities to reason, plan, solve problems, think abstractly, comprehend ideas and language, and learn.

  It's not that I'm so smart; it's just that I stay with problems longer.
  Albert Einstein

Affirming what we view as valuable at the outset requires that we clarify what we mean when we say we value some one or some idea. In essence, what we are doing is clarifying what we value, and why we value it. A sevenfold process describing the guidelines of the values clarification approach was formulated by Simon et al. (1995).

- choosing from alternatives;

- choosing freely;

- prizing one's choice;

- affirming one's choice;

- acting upon one's choice; and

- acting repeatedly, over time.

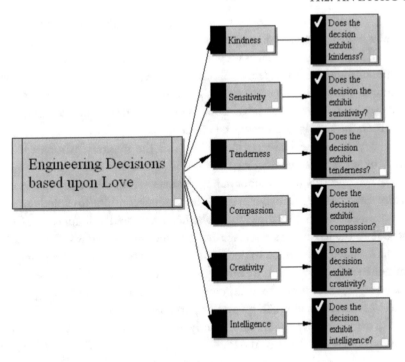

Figure 11.3: Seven elements of engineering decisions based upon love.

Additional theorists providing support for the values clarification approach include Asch (1952) and G. Murphy (1958).

The last category of love, *philia*, challenges us to see the world in a different way, in the words of Thomas Berry (2006) as a *collection of subjects*. Berry's most famous quotation is:

The Universe and thus the Earth is a communion of subjects, not a collection of objects.

By communion, Berry was referring to intimacy or a feeling of emotional closeness, a connection, especially one in which something is communicated or shared. The shift from object to subject[1] is also profound. An object is something visible or tangible; something that can be seen or touched, a focus of somebody's attention or emotion; or a goal or purpose. By subject, the reference is to the essential nature or substance of something as distinguished from its attributes. In other

---

[1] In Physicist Conception of Nature, Werner Heisenberg's underlines the fundamental chance in the status of subject/object relationship, brought about by the quantum theory (and the Copenhagen interpretation of it). Newtonian physics has a clear-cut distinction between object and subject. When an XIX century physicist was approaching the study of the nature, he was hoping to unveil the law of it; the subject of his study was nature "itself." After quantum physics this is no longer possible - there is no "nature itself." The process of observation for ever changes the observed. The observer and the observed are interacting. Heisenberg writes: "We can no longer speak of the behavior of the particle independently of the process of observation." The laws we formulate are not about the nature itself, but about our knowledge of it.

words, borrowing from Buddhism,[2] the essential nature, the Buddha nature, is taught to be a truly real, but internally hidden, eternal potency or immortal element within the purest depths of the mind, present in all sentient beings. When we practice our profession of engineering, it is important that we view humanity and the ecosystem as part of an undivisable whole. Berry takes this one step farther. According to Berry, our new community is a very special one, that is, it is one in which the various elements are bound together as subjects having interests rather than one in which some have interests while others are simply resources to be utilized.

In our view, viewing the Universe including both the natural environment and the poor as a communion of subjects rather than a collection of objects has important, even revolutionary significance for the engineering profession. First, it eliminates from the outset that we ever again can remain aloof from the consequences of our projects. Landmines continue to explode long after the end of hostilities in combat zones across the globe. Polar bears are rapidly disappearing from the Arctic regions in part due to the technologies we continue to produce. The poor in New Orleans suffered beyond our understanding in part due to decisions we as engineers and engineering organizations made and continue to routinely make. If we can begin to see the connection we have with the health of the Arctic ecosystem and thus with the well-being of the polar bears, recognizing all that we share; they like the rest of nature have much greater importance when we are formulating our criteria whereby we make decisions. Even more importantly, if we can begin to view the poor, rather they live in the 9th Ward of New Orleans or the Pine Ridge Reservation in South Dakota, as connected to us and as possessing an entire spectrum of potentialities and possibilities then too our criteria for decision making as engineers is broadened importantly. Those potentialities and possibilities are as important to the ongoing dynamic process of creation in the Universe as those that reside within us as each of us plays an integral role in the communion of subjects.

In engineering, we often speak of development. Far too often, it seems that the model used in engineering is linked solely to profit making. The ultimate goal is economic growth with no interest in peace, social or environmental justice or wealth distribution. Such a model ignores inequalities, has contempt for the arts and literature, promotes group think, needs docile practitioners and de-emphasizes critical thinking. I would like to offer a different paradigm for engineering, one which has as its priority the development of not only the human spirit but also the rest of the natural world. Using such a paradigm, each and every being matters, groups are disaggregated into individuals and equal respect exists for each individual. Ultimately, the goal of such a profession would be to enable each of us to transcend our own particular situations and imagine a global society which is based upon equality and on love. Key elements of our profession based upon love would include the capacity for true, rigorous critical thought, the development of a culture in which individual dissent is honored and revered, and in which each of us considers our self a citizen of the Earth. Lastly, an engineering based upon love would enable each of us to develop our own individual narrative of

---

[2] Buddhism is a dharma, non-theistic religion, a philosophy, and a system of psychology. Buddhism is also known in Sanskrit or Pail, the main ancient languages of Buddhists, as Buddha Dharma which means the teachings of "the Awakened One."

moral imagination, that is, to develop the ability to be in another's shoes, to cultivate our inner eye of seeing and knowing and to overcome the blindness that we have all become far too accustomed.

Challenge Box: Summarize in your own words what is meant by an engineering ethic based on love. Suppose you were challenged to write down such an ethic for engineering in less than 100 words, what would you write?

What similarities would there be between ethical approach versus other more traditional approaches we have already taken? What differences?

Let's go back to our trusted trolley problem. What would this ethical point of view point to for both scenarios? Explain. Are the recommendations different than what you concluded earlier?

CHAPTER 12

Case Study Application

We shall conclude with several different case studies, each representative of the kinds of real-world decision each of you, as engineers, may confront in your upcoming professional careers. The present text has laid out a range of options whereby you can carefully consider the cases in light of applied ethical theories ranging from those traditionally used in engineering to new ideas based on developments in moral reasoning to include issues such as freedom, chaos, a morally deep world view, and globalism, and ultimately, one based on love. After the case is described, some possible suggestive questions will be provided as a means for you to begin your own, personal reflection on the various cases and your responses as a future engineering professional.

12.1 THE NEXT GENERATION OF LAND MINES

Part I.

Ms. Jane Enaj is a project manager at a multinational corporation which as just been awarded a contract to develop and produce the next generation land mine. She is also a member of the Design Review Committee. The committee's responsibilities include reviewing and approving design changes, procedural changes and submitting performance reports to various U.S. Department of Defense agencies with recommendations.

Today, Jane finds herself in a difficult situation. DRC is meeting to finalize recommendation concerning the new land mine. It offers significant improvements for the U.S. military as it will self-detonate in a set period of time thereby reducing the risk of having unexploded ordinance remain on the battlefield long after the actual conflict has ended. This new design should reduce the number of injuries and deaths to innocent civilians by a large amount and is set for deployment in the Middle East. Yet innocent citizens including children and the elderly will still be maimed and/or die. In addition, unsuspecting pets, domestic animals and wildlife will also likely suffer.

The unit has returned after a two month trail run in various environments including U.S desert southwest, and island possessions in the Pacific. After extensive engineering analysis by engineers, it has been concluded that the new design falls slightly above the minimum requirement set by the design specifications. The test and analysis results were both promising and disappointing as approximately 50% of the mines self-destructed. This still would represent a significant improvement over the use of the land mines in the present U.S. arsenal.

Seven members of DRC are present, enough for a quorum. Ms. Enaj is the least senior member present. From the outset of the meeting, committee chair Mr. Senior has made it clear that it is important to act quickly since any delay will cost the company, and a lot of money. "A redesign," he says, "might take several months. If we don't approve this, we may be facing a multi-million dollar loss in revenues. We have met the design specifications. What do you think?" Mr. Smith and Mr. Jones immediately concur. Mr. Senior then says, "Well, if no one sees any problems here, let's go with it." There is a moment of silence. Suppose you were Ms. Enaj. What would you recommend?

Part II.

Right after the design meeting a breakthrough was made in the design of the actual explosive material used in the mine. For the same amount of explosives material used, the magnitude of the explosion is doubled both in intensity and radius of impact. The DRC calls a second meeting to consider what impact the new explosives material may have on their final recommendations. Again playing the role of Ms. Enaj, what would you recommend and/or do?

Part III.

After the land mine has been put into use by American forces, the Chinese and Pakistani governments submit orders for significantly large number of the new devices. China and Pakistan are in turn is known to sell similar weapon systems to various regimes throughout the world. Again playing the role of Ms. Enaj, what would you recommend and/or do?

12.2   ADVANCES IN AUTOMOTIVE TECHNOLOGIES IN THE DEVELOPING WORLD

Part I.

Mr. Sam Mas is a project engineer for the electronics branch of a multinational automotive company. Mr. Mas' design team has recently designed, produced and tested a new control module for possible use in a mass produced economy car. The new module has the potential for reducing the manufacturing cost as well as increasing the energy efficiency of the car. The new design is ideally suited as part of the company's strategy to increase sales in under-developed nations throughout the world. The new car has the potential to revolutionize transportation in regions of the globe where travel is still quite expensive. The company's board of directors is meeting in order to decide on the future for this new module. The plans are to produce the first version of the module in northern California and then subsequently to transfer production to Bangladesh to reduce labor and manufacturing costs. Mr. Mas is asked to provide his recommendations to the board as they consider implementation of the new system. Acting in the role of Mr. Mas, what would you recommendations to the board be?

Part II.

The new module has been put into production with the production facilities transferred to Bangladesh as planned. The sale of automobiles has increased dramatically in less affluent countries. The production facility in the U.S. has been abandoned with the resultant dislocation of the workers involved. The neighborhood in which the production facility has been located has seen a rapid

Next Generation of Landmines

Utilitarianism	Rights of Persons
□ What are all the available options for Ms. Enaj *concerning* the design review? □ What are the costs of delaying the introduction of the landmine? □ Who will benefit the most from its introduction? □ What costs are associated with increased explosive capabilities? □ Who will benefit the most from its increased explosive capabilities? □ What decisions maximize the good? □ What are the costs associated with the selling of the weapons system to China and Pakistan?	□ What are the effects of the improved land mine on the combatants and on civilians? □ Would we be willing to serve as soldiers under such an arrangement? As civilians? □ Would we be able to accept the consequences of the new landmine on ourselves as soldiers? □ Would we be able to accept the consequences of the new landmine on ourselves as innocent civilians?
Virtue	Freedom
□ Are the rights of all involved in the landmine production and use given their just due? □ Does the new design and its implementation demonstrate self-control and/or discipline? □ Does the design and use of the new landmine design require moral strength and/or courage? □ Does its design and use signify a pursuit of good according to reason? □ Or is it simply a manifestation of our fears?	□ In designing and using the new landmine design are we cognizant of the responsibility we have to all of humankind? □ Are we working for the freedom of all of humanity? □ Are we aware of the context of relationships that exist? □ Are we cognizant of the obligations we have to ourselves as well as to others? □ Are we aware of our individuality and take it into account?

Figure 12.1: Next generation of landmines: suggested reflective questions.

transformation away from the original native culture. The company's board of director is meeting again to consider even greater increases in control module production. Mr. Mas is invited to provide engineering's perspective. Acting in the role of Mr. Mas, what would you recommendations to the board be?

Next Generation of Landmines

Chaos	Morally Deep World
□ Are we taking into account in our landmine design the integrity of the biotic community? □ Are we considering the stability of the same community as well as the stability of the cultures involved? □ Are we cognizant of the beauty of the local environment, different cultures including our own? □ Are we enriching the potential experiences of the various elements within the communities? □ Are we respecting and promoting diversity of not only species but also thought and custom? □ Are we arbitrarily seeking to restrain change? □ Are we comfortable with uncertainty?	□ For whom are we designing the landmine? □ Whose interests does it serve? □ Have we identified the integral community for the design? □ Who is included in the integral community? □ What other elements are involved in the integral community? □ Have we called on our vast array of creativity as well as analytical skills? □ have we considered the possible negative consequences involved in the mine design? □ Has the pain and suffering in the world increased, decreased, remained the same or unknown as our design is put into theaters around the world?

Globalism	Love
□ Do we consider the global ethic when we design the new landmine? □ Are we cognizant of the interdependence of all life and the planet? □ Are we promoting a more equitable order with the landmine use? □ Is there a chance that more will benefit as a result of its use? □ Or does it serve the self-serving interests of a few? □ Have we tried in many different ways to fully understand the basis for the conflict itself? □ Are we treating those with whom we come into contact with our devices in the same way we would wish to be treated?	□ Does the new design exhibit any elements of kindness? □ To those who must use it? □ To those who will ultimately be maimed or killed? □ Is it sensitive to the needs of not only the combatants but also the innocent civilians and the ecosystem? □ Does the device foster a sense of protectiveness? For whom? □ As designers, have we demonstrated compassion for whom will be affected by our device? □ Have we been as creative as we can in its design, execution, implementation? □ Have we ben willing to think outside the box, to use our mind to its fullest capabilities?

Figure 12.2: Next generation of landmines: suggested reflective questions.

12.3   LEVEES, COASTLANDS, AND SAFETY

Part I.

Dr. Zeeman, a career governmental employee, is the chief engineering officer for the government agency overseeing the design and reconstruction of the levees in and around the New Orleans

Advances in Automotive Technology

Utilitarianism	Rights of Persons
□ What are all the available options for Mr. Mas concerning the design review? □ What are the costs of delaying the introduction of the module? □ Who will benefit the most from its introduction? □ What costs are associated with increased sales of autos? □ Who will benefit the most from its increased sales and use? □ What decisions maximize the good? □ What are the costs associated with moving the plant to Bangladesh?	□ What are the effects of increased use of autos? □ Would we be willing to accept more autos in our towns? Our country?? □ Have we considered the impact of changes to transportation to the native culture(s)? □ Would we be able to accept the consequences of the new landmine on ourselves as innocent civilians? □ Have we considered the effects of job relocation on those originally employed in the U.S.?
Virtue	Freedom
□ Are the rights of all involved in the control production and use given their just due? □ Does the new design and its implementation demonstrate self-control and/or discipline? □ Does the design and use of the new control module require moral strength and/or courage? □ Does its design and use signify a pursuit of good according to reason? □ Or is it simply a manifestation of our selfishness or greed?	□ In designing and using the new module design are we cognizant of the responsibility we have to all of humankind? □ Are we working for the freedom of all of humanity? □ Are we aware of the context of relationships that exist? Among the workers in the U.S.? Among the workers in Bangladesh? □ Are we cognizant of the obligations we have to ourselves as well as to others? □ Are we aware of our individuality and take it in to account?

Figure 12.3: Automotive technology: suggested reflective questions.

area. His agency has performed an in-depth and careful analysis of the failure of the levees as a result of Hurricane Katrina. The final report provides a design for three different scenarios: a Category 3, a Category 4 and a Category 5 strength hurricane striking the metropolitan area of New Orleans. The new design focuses on strengthening the levee system as it presently exists. A meeting of Dr. Zeeman as well as appropriate governmental officials is called with the purpose to finalize plans for redesign. A budget crisis has reduced the amount of funds available for the project with the result

Advances in Automotive Technology

Chaos	Morally Deep World
□ Are we considering the integrity of the biotic community with new auto production and use? □ Are we considering the stability of the same community as well as the stability of the cultures involved with the increased use? □ Are we cognizant of the beauty of the local environment, different cultures including our own? □ Are we enriching the potential experiences of the various elements within the communities? □ Are we respecting and promoting diversity of not only species but also thought and custom? □ Are cognizant of the various societies, their customs, and their histories?	□ For whom are we designing the control module? □ Whose interests does it serve? □ Have we identified the integral community for the design? □ Who is included in the integral community? □ What other elements are involved in the integral community? □ Have we called on our vast array of creativity as well as analytical skills? □ Have we considered the possible negative consequences involved in the module design? □ Has the pain and suffering in the world increased, decreased, remained the same or unknown as our design is put into use around the world?
Globalism	Love
□ Do we consider the global ethic when we increase use of cars? □ Are we cognizant of the interdependence of all life and the planet? □ Are we promoting a more equitable order with more auto use? □ Is there a chance that more will benefit as a result of its use? □ Or does it serve the self-serving interests of a few? □ Have we tried in many different ways to fully understand the basis for the conflict itself? □ Are we treating those whose lives are changing due to the design, production and relocation of production in the same way we would wish to be treated?	□ Does the new design exhibit any elements of kindness? □ To those who must use it? □ To those who will ultimately be affected by it? □ Is it sensitive to the needs of not only the producers, the users as well as the bystanders? □ Does the device foster a sense of protectiveness? For whom? □ As designers, have we demonstrated compassion for whom will be affected by our device? □ Have we been as creative as we can in its design, execution, implementation? □ Have we been willing to think outside the box, to use our mind to its fullest capabilities?

Figure 12.4: Advances in automotive technology: suggested reflective questions.

being. Dr. Zeeman is asked to sign off the plan to insure that the levees can withstand a Category 3 hurricane. Suppose you were the take the place of Dr. Zeeman at the meeting, would you do?

Part II.

A study from a nearby land grant university points to the accelerating erosion of the Louisiana coastland as a result of the levees system. This study comes to your attention immediately after the planning meeting on the Category 3 design implementation. The loss of coastal wetlands increases the risk of damage from subsequent hurricanes as well as the destruction of the local ecosystem with loss of plants, wildlife and income for those residents who make their income from the area. Again acting in the role of Dr. Zeeman, what would you do?

Part II.

The levee reconstruction goes ahead as planned with no attention paid to the issue of population evacuation. The plans in place have not change since Hurricane Katrina. Your agency is not responsible for developing an evacuation plan. A local newspaper consults you on a story involving post-Katrina readiness and safety. How would you respond to the reporter's questions?

12.4 NATIONAL PARKS, DANGEROUS DRIVERS, AND THE REMOVAL OF TREES

Part I.

Isabella Arbole is the Chair of the Road Safety Commission for the Rocky Mountain National Park (RSCRMNP). Her agency has primary responsibility for maintaining the safety of all roads within the confines of the park. Visitors to the park have increased by 200% in the past 10 years. This has resulted in increased traffic flow on both primary and secondary roads in the area. Alpine Drive, still a two lane road, has more than tripled its traffic flow during this period.

For each of the past 10 years at least one person per year has suffered a fatal automobile accident by crashing into trees closely aligned along a 5 mile stretch of Alpine Drive. Many other accidents have also occurred, causing serious injuries, wrecked cars, and damaged trees. Some of the trees are quite close to the pavement. Two law suits have been filed against the park for not maintaining sufficient road safety. Both cases were decided in the government's favor as not enough evidence was submitted to prove the government's responsibility for the crashes as in each case the drivers were traveling way above the posted speed limit and the road conditions were dangerous due to freak storms.

Other members of RSCRMNP have been pressing Ms. Arbole to propose a solution to the traffic problem on Alpine Drive. They are concerned about safety, as well as law suits that may some day go against RSCRMNP. Suppose you are acting as Ms. Arbole's engineering expert, what would you recommend?

Part II.

A significant increase in the elk population occurs in the park. The increase while not threatening the balance of the ecosystem does pose a significant problem for safety on the roadways. The numbers of deaths as a result of automobile-elk crashes has increased from less than one per year to on average three per year. The other members of RSCRMNP have been pressing Ms. Arbole to propose a solution. They are again concerned about safety, as well as law suits that may some day go against RSCRMNP. Suppose once again you are acting as Ms. Arbole's engineering expert, what would you recommend?

Levees, Coast lands and Safety

Utilitarianism	Rights of Persons
□ What are all the available options for the levee redesign? □ What are the costs associated with the design? □ Who will benefit the most from its completion? □ What costs are associated with increased coastal erosion? □ Who will benefit from the status quo? □ What decisions maximize the good? □ What are the evacuation considerations and costs associated with the new levee system?	□ What are the effects of the new design on past, present and future residents? □ Would we be willing to move/live in New Orleans under such an arrangement? □ Would we be able to accept the consequences of the new design? □ Would we be able to accept the consequences of the new design on ourselves as innocent civilians? □ Would we be willing to live in New Orleans realizing that we may not be able to evacuate?
Virtue	Freedom
□ Are the rights of all involved in the levee redesign given their just due? □ Does the new design and its implementation demonstrate self-control and/or discipline? □ Does the design require evidence of moral strength and/or courage? □ Does the design and use signify a pursuit of good according to reason? □ Or is it simply a manifestation of our fears? Or biases? Or prejudices?	□ In designing and using the new levee design are we cognizant of the responsibility we have to all of humankind? □ Are we working for the freedom of all of humanity? □ Are we aware of the context of relationships that exist? □ Are we cognizant of the obligations we have to ourselves as well as to others? □ Are we aware of our individuality and take it in to account?

Figure 12.5: Levees, coastlands and safety: suggested reflective questions.

Levees, Coast lands and Safety

Chaos	Morally Deep World
▫ Are we considering the integrity of the biotic community in New Orleans? ▫ Are we considering the stability of the same community as well as the stability of the cultures involved with the increased use? ▫ Are we cognizant of the beauty of New Orleans? ▫ Are we enriching the potential experiences of the various elements within the communities? ▫ Are we respecting and promoting diversity of not only species but also thought and custom? ▫ Are cognizant of the various societies, their customs, and their histories?	▫ For whom are we redesigning the levee system? ▫ Whose interests does it serve? ▫ Have we identified the integral community for the design? ▫ Who is included in the integral community? ▫ What other elements are involved in the integral community? ▫ Have we called on our vast array of creativity as well as analytical skills? ▫ Have we considered the possible negative consequences involved in proposed levee redesign? ▫ Has the pain and suffering in the world increased, decreased, remained the same or unknown for the various elements of the integral community?
Globalism	Love
▫ Do we consider the global ethic when we redo the levee system? ▫ Are we cognizant of the interdependence of all life and the planet? ▫ Are we promoting a more equitable order with more auto use? ▫ Is there a chance that more will benefit as a result of its use? ▫ Or does it serve the self-serving interests of a few? ▫ Have we tried in many different ways to fully understand the implication of the design and its omissions? ▫ Are we treating those whose lives are changing due to the design in the same way we would wish to be treated?	▫ Does the new design exhibit any elements of kindness? ▫ To those who must use it? ▫ To those who will ultimately be affected by it? ▫ Is it sensitive to the needs of not only the producers, the users as well as the bystanders? ▫ Does levee system provide a sense of protectiveness? For whom? ▫ As designers, have we demonstrated compassion for whom will be affected by our design? ▫ Have we been as creative as we can in its design, execution, implementation? ▫ Have we been willing to think outside the box, to use our mind to its fullest capabilities?

Figure 12.6: Levees, coastlands and safety: suggested reflective questions.

National Parks, Dangerous Drivers and Trees

Utilitarianism	Rights of Persons
□ What are all the available options for Ms. Arbole concerning the design ? □ What are the costs of new roads and tree removal in the park? Of delaying the new design? □ Who will benefit the most from its introduction? □ What costs are associated with increased use? □ Who will benefit the most from its increased use? □ What decisions maximize the good? □ What are the costs associated with the increased use, tree removal, elk harvesting?	□ What are the effects of the improved roads on the visitors? On local residents? □ Would we be willing to serve as live under such an arrangement? A □ Would we be able to accept the consequences of the increased visitation, use and speed of travel? □ Would we be able to accept the consequences of the new system on ourselves, our communities, and our local ecosystems?
Virtue	Freedom
□ Are the rights of all involved in the road design their just due? □ Does the new design and its implementation demonstrate self-control and/or discipline? □ Does the design require moral strength and/or courage? □ Does its design and use signify a pursuit of good according to reason? □ Or is it simply a manifestation of our fears? □ Or is it simply a manifestation of our greed? □ Or is it simply a manifestation of our narcissism ?	□ In designing and using the transportation design are we cognizant of the responsibility we have to all of humankind? □ Are we working for the freedom of all of humanity? □ Are we aware of the context of relationships that exist? □ Are we cognizant of the obligations we have to ourselves as well as to others? □ Are we aware of our individuality and take it in to account?

Figure 12.7: National parks, dangerous drivers, and the removal of trees: suggested reflective questions.

National Parks, Dangerous Drivers and Trees

Chaos	Morally Deep World
□ Are we considering the integrity of the biotic community in Rocky Mountain National Park? □ Are we considering the stability of park as well as the stability of the cultures involved with the increased use? □ Are we cognizant of the beauty of park? □ Are we enriching the potential experiences of the various elements within the park? □ Are we respecting and promoting diversity of not only species but also thought and custom? □ Are cognizant of the various ecosystems, societies, their customs, and their histories?	□ For whom are we redesigning the road system in the park? □ Whose interests does it serve? □ Have we identified the integral community for the design? □ Who is included in the integral community? □ What other elements are involved in the integral community? □ Have we called on our vast array of creativity as well as analytical skills? □ Have we considered the possible negative consequences involved in proposed road redesign? □ Has the pain and suffering in the world increased, decreased, remained the same or unknown for the various elements of the integral community?
Globalism	Love
□ Do we consider the global ethic when we redo the road system in the park? □ Are we cognizant of the interdependence of all life in the park and the planet? □ Are we promoting a more equitable order? □ Is there a chance that more will benefit as a result of more and safer roads? □ Or does it serve the self-serving interests of a few? □ Have we tried in many different ways to fully understand the implication of the design and its omissions? □ Are we treating those whose lives are changing due to the design in the same way we would wish to be treated?	□ Does the new design exhibit any elements of kindness? □ To those who must use it? □ To those who will ultimately be affected by it? The trees? The elk? □ Is it sensitive to the needs of all? □ Does road system provide a sense of protectiveness? For whom? □ As designers, have we demonstrated compassion for whom will be affected by our design? □ Have we been as creative as we can in its design, execution, implementation? □ Have we been willing to think outside the box, to use our mind to its fullest capabilities?

Figure 12.8: National parks, dangerous drivers and the removal of trees: suggested reflective questions.

CHAPTER 13

Final Thoughts

I am the voice crying in the wilderness...the voice of Christ in the desert of this island...[saying that] you are all in mortal sin...on account of the cruelty and tyranny with which you use these innocent people. Are these not men? Have they not rational souls? Must not you love them as you love yourselves?

Antonio de Montesinos (1987).

The aim of the present text is to provide each of you with the skills to be able to confront the challenges that awaits us as professional engineers in the 21st century. The world is becoming ever more complicated, ever more interconnected. It is our belief that it only serves to help us if we are aware of a wide range of possible approaches we may take when confronted with the need to make decisions today and in the future. We are challenging you and the discipline of engineering to continue to embody an ethical approach to issues of safety, and professionalism and the like but also to take a much broader view of our macro-ethical responsibilities. That is why we have provided you with new tools of applied ethical analysis to include notions of freedom, of chaos, of a morally deep world view, of globalism and ultimately of love. This is an important shift in the engineering profession and in our view one that is desperately needed.

In many ways, the shift that we are calling for is remarkably similar to an impassioned call for a radical shift in consciousness in Western thought that occurred almost 500 years earlier in Spain. It is a debate that brought to the forefront many of the same issues we are dealing with now in the 21st century. The confrontation was known as the Valladolid debate and it concerned the treatment of natives of the New World. Held in the city of Valladolid in Spain, the debate featured the two main attitudes taken in Spain towards the conquests of the New World. On one side, Dominican monk Bartolome de Las Casas argued that the natives were free men in the natural order and deserved the same treatment as others, according to Catholic theology. Las Casas was opposed by fellow Dominican monk, Sepulveda, who insisted the Indians were natural slaves, and therefore reducing them to slavery or serfdom was in accordance with Catholic theology and natural law.

The debate focused on questions such as: What are just wars? When is violence justified? What are the responsibilities of the developed world towards the under-developed world? What does it mean to be a human being? What rights do all human being possess by virtue of being human? What rights if any does the planet possess?

Sadly, no positive outcome came out of the debate; no realistic solution could have resulted, for the debate was carried out in too theoretical a framework. Both sides, determined to prove or disprove the legality of war as a means of conversion, adamantly stuck to their respective writings, and thus failed to reach a realistic and concrete compromise. Not surprisingly, the debate failed to materialize into palpable benefits for the Indians.

Though concrete steps to alleviate the suffering of the Indians did not result as a result of the debate, that event signaled the beginning of a long journey towards greater awareness of the plight of the Indians but also of other groups such as those ensnared in the African slave trade. Consciousness was raised. It was a beginning of a conversation, a conversation that continues to this day. Our hope is that in some small way our work will result in the beginnings of a conversation and a similar awakening of consciousness in the profession of engineering today as we move ever farther into the 21st century.

Bibliography

AIChE Code of Ethics, 2008.
  http://www.stuorg.iastate.edu/aiche/ethics.html

Allenby, B., *Micro and Macro Ethics for an Anthropogenic Earth*, Professional Ethics Report, Vol. 18, No. 2, Spring 2005.

Amartya Sen, *Development as Freedom*, Anchor Press, 2005.

Annas, J., *Virtue Ethics*,
  http://www.u.arizona.edu/~jannas/forth/coppvirtue.htm

ASCE Code of Ethics, 2008.
  http://www.asce.org/inside/codeofethics.cfm

Asch, S., *Social psychology*. Englewood Cliffs, NJ: Prentice-Hall, 1952.

ASME Code of Ethics, 2008.
  http://courses.cs.vt.edu/~cs3604/lib/WorldCodes/ASME.html

Asprey, R., *The Rise of Napoleon Buonoparte*, Basic Books, 2002.

Aurelius, Marcus, Hicks, D., and Hicks, C. Scot, *The Emperor's Handbook: A New Translation of the Meditations*, Scribner, 2002.

Austin, J., *The Province of Jurisprudence Determined*, Cambridge University Press, 1995.

Berry, T., *The Dream of the Earth*, Sierra Club, October 2006.

Chomsky, N., "Education is Ignorance," excerpted from *Class Warfare*, 1995, pp. 19–23, 27–31.

Common Dream.org, 0000.
  http://www.commondreams.org/views05/0907-26.htm

Cook, M., *The Moral Warrior: Ethics and Service in the U.S. Military*, Albany, NY, State University of New York Press, 2004.

Dalai Lama and Cutler, H., *The Art of Happiness: A Handbook for Living*, Riverhead Press, 1998.

Daly, H., Population, Migration, and Globalization, Worldwatch Magazine,
  http://www.publicpolicy.umd.edu/faculty/daly/WW%20rev%20pop,migr,
  glob%20copy%201.pdf

Davis, M., *Ethics and the University*, Routledge, 1999.

Davis, M., *Thinking Like an Engineer: Studies in the Ethics of a Profession,* Oxford University Press, 1998.

Decker, 1995.

Ellis, J., *Colony Collapse Disorder (CCD) in Honey Bees,*
http://pestalert.ifas.ufl.edu/Colony_Collapse_Disorder.htm

Friedman, T., *The World is Flat 3.0*, Picador Press, 2007.

Garrett, J., *Amartya Sen's Ethics of Substantial Freedom*,
http://www.wku.edu/~jan.garrett/ethics/senethic.htm

Gelinas, N., *Katrina's Real Lesson*, City Journal, Vol. 17., No. 1, Winter 2007.

Graves, K., Golden Rule by Sextus, *The World's Sixteen Crucified Saviors*, Colby and Rich, 1876.

Harris, C., Pritchard, M., and Rabins, M., *Engineering Ethics: Concepts and Cases*, Wadsworth Pub., July 2008.

Herkert, J., *Engineering Ethics and Climate Change*, ASEE 2008, Pittsburgh, Pa., June 2008.

Hinman, L.,
http://ethics.sandiego.edu/presentations/Theory/Utilitarianism///
index_files/v3_document.html

Wikipedia, 2008.
http://en.wikipedia.org/wiki/Brian_Swimme

Wikipedia, 2008.
http://en.wikipedia.org/wiki/Profession

Holmes, 2002.
http://users.ece.utexas.edu/~holmes/Teaching/EE302/Slides/
UnitOne/sld002.htm

Brainy Quotes, 2008.
http://www.brainyquote.com/quotes/quotes/e/eowilson164833.html

NOAA, 2005.
http://www.katrina.noaa.gov/satellite/images/katrina-08-29-2005-1415z.jpg

Quoteworld, 2008.
http://www.quoteworld.org/category/nature/author/ralph-waldo-emerson

MFGB, 0000.
http://www.santafe.edu/~gmk/MFGB/node2.html

Hurricane Katrina Damage- Louisiana, Adjutant General's Department State of Kansas,
http://www.accesskansas.org/ksadjutantgeneral/Disaster-Emergency/2005/
Hurricane%20Katrina/Damage/Damage%20-%20Louisiana.htm

IEE Code of Ethics, 2008.
http://www.iienet2.org/Landing.aspx?ID=299

IEEE Code of Ethics, 2008.
http://www.ieee.org/portal/pages/iportals/aboutus/ethics/code.html

Jagger, B., Cluster Bombs and Teddy Bears, Telegraph.co.uk., May 11, 2007,
http://www.telegraph.co.uk/opinion/main.jhtml?xml=/opinion/2007/11/05/
do0502.xml

Jefferson, T., to Maria Cosway, 1786. ME 5:443.

Johnson, L., 1991.

Johnson, L., *A Morally Deep World*, Cambridge University Press, 1993.

Kant, I., 'Preface.' In *The Metaphysical Elements of Ethics*. Translated by Thomas Kingsmill Abbott, 1780.

Kashner, A., Professional Ethics and Collective Personal Autonomy, *Ethical Perspectives*, 12/1, March 2005, pp. 67–97.

King, M.L., and Washington, J.M., *A Testament of Hope: The Essential Writings and Speeches of Martin Luther King*, Harper One, 1990.

Landmines: Africa's Stake, Global Initiatives,
http://www.africaaction.org/bp/lmineall.htm

Leopold, A., *Sand County Almanac*, Oxford University Press, 1968.

Liaschenko, J., Oguz, N. and Brunnquell, D., *Critique of the Tragic Case Method in Ethics Education*, *Journal of Medical Ethics*, 32(11):672.

Luke 6, 27–32, English Standard Version Bible.

Marsh, G.P., *Man and Nature*, Kessenger Publishing, 2008.

Mayer-Kress, G., *What is Chaos?*, Sat Apr 22, 21:04:59 MDT, 1995,
http://www.santafe.edu/~gmk/MFGB/node2.html

Metairie, Louisiana, Wikipedia,
http://en.wikipedia.org/wiki/Metairie%2C_Louisiana

Mills, J.S., *On Liberty and Other Essays*, Oxford University Press, 2008.

Murphy, G., *Human Potentialities.* New York: Basic Books, 1958.

Nature editorial, 2003, "Welcome to the Anthropocene," *Nature* 424:709.

News Hounds,
http://www.newshounds.us/2005/08/31/when_the_levee_breaks.php

NSPE Code of Ethics, 2008.
http://www.nspe.org/Ethics/CodeofEthics/index.html

Nussbaum, M.C., *Sex and Social Justice*, Oxford University Press, 1999.

Pacific Disaster management Information, Network 2004.

Panch, T., *Enduring Effects of War: Health in Iraq 2004*, Medact, London, UK, 2004.

Petroski, H., *To Engineer is Human: the Role of Failure in Successful Design*, St Martins Press, 1985.

Pojman, L.P., and Fieser, J., *Ethics: Discovering Right and Wrong*, Wadsworth Cengage Learning, Wadsworth: Belmont, Ca., 2009.

Random House Webster's Unabridged Dictionary, Random House Reference, 2005.

ScienceDaily, Apr. 23, 2007.

Sherman, W.T., *Memoirs*, Penguin Classics, 2000.

Simon, S.B., Howe, L.W., and Kirschenbaum, H., Values Clarification, Grand Central Publishing; Revised edition (September 1, 1995).

Singer, P., *In Defense of Animals: The Second Wave*, Wiley-Blackwell, 2005.

Singh, G.B., *Gandhi: Behind the Mask of Divinity*, Prometheus Books, 2004.

Smith, H.W., *Aeschylus*, Harvard University Press, 1970.

Stanford Encyclopedia of Philosophy,
http://plato.stanford.edu/entries/rights/

Stewart, I., 1998.

Stewart, I., *Does God Play Dice*, Wiley Blackwell, 2002.

Stockdale, J., *A Vietnam Experience: 10 Years of Reflection*, Hoover Institution Press, 2004.

Tesconi, T., "California Grape Harvest 2000," *The Press Democrat*, October 26, 2000.

The Ten Commandments, Catholic Encyclopedia,
http://www.newadvent.org/cathen/04153a.htm

The Universality of the Golden Rule in the World Religions,
www.teachingvalues.com/goldenrule.html

Twain, M., "To The Person Sitting in Darkness," Mark Twain's Weapons of Satire: Anti-Imperialist Writings on the Philippine-American War, Syracuse University Press, 1992.

Ulam, S., on Chaos, 0000.
http://www.santafe.edu/~gmk/MFGB/node2.html

von Clausewitz, Carl, *On War*, Eco Library, 2007.

Warner, C., "All Mankind Is One:" The Libertarian Tradition In Sixteenth Century Spain. *The Journal of Libertarian Studies*, 8, 2:293–309, 1987.

White House: U.S. will ban some land mines, CNN.com, February 27, 2004,
http://www.cnn.com/2004/US/02/27/bush.landmines.ap/

Worldwatch Institute, *State of the World 2007*, New York: W.W. Norton, 2007.

Printed in the United States
by Baker & Taylor Publisher Services